GENETIC ENGINEERING: The Hazards
VEDIC ENGINEERING: The Solutions

Health—Agriculture—The Environment

Maharishi's Vedic Engineering—
Comprehensive, life-supporting solutions
Genetic Engineering—Partial, short-term
fixes with damaging side effects

•

JOHN FAGAN, Ph.D.

Maharishi International University Press
Fairfield, Iowa USA

© 1995 by Maharishi International University. All rights reserved. No part of this publication may be reproduced, stored in a retrieval system, or transmitted in any form or by any means, electronic, mechanical, photocopying, recording, or otherwise, without prior written permission of the publisher. Transcendental Meditation®, TM®, Science of Creative Intelligence®, SCI®, TM-Sidhi®, Vedic Science, Maharishi®, Maharishi International University®, MIU®, and the MIU logo® are service marks used under license by Maharishi International University and the World Plan Executive Council–United States, a nonprofit educational organization. Maharishi Ayur-Veda is a service mark used under license by Maharishi Ayur-Veda Foundation. Printed in the United States of America.

ISBN 0-923569-18-9

"Genetic engineering breeds costly protest," Thursday, November 17, 1994, and "A scientist's qualms," Monday, November 21, 1994, reprinted courtesy of *The Washington Post*.

"Cancer researcher returns grant," Friday, November 25, 1994, reprinted courtesy of *Science*.

"Biologist returns US grants to protest genetic research," Wednesday, November 16, 1994, reprinted courtesy of *The Boston Globe*.

"Scientist returning grant, opposes genetic engineering," Friday, November 18, 1994, reprinted courtesy of *Chicago Tribune*.

"Gene scientist, with warning, spurns grant," Fri./Sat./Sun., Nov.18–20, 1994, reprinted courtesy of *USA Today*.

"Researcher returns grant," Friday, November 29, 1994, reprinted courtesy of *Hindustan Times*.

"Scientist gives back research funds," Saturday, November 19, 1994, reprinted courtesy of *The China Post*.

"Researcher returns NIH funding in protest," December 19, 1994, reprinted courtesy of *Chemical & Engineering News*.

"When science gets ahead of ethics," and "Genetic engineering's sinister side," Sunday, December 11, 1994, reprinted courtesy of *The Des Moines Register*.

FOR INFORMATION, PLEASE CONTACT:
Maharishi International University
1000 North Fourth Street
Fairfield, Iowa 52557
or
Maharishi Vedic University, International
Station 24
6063 NP
Vlodrop, Netherlands

IN GRATITUDE TO MAHARISHI MAHESH YOGI whose Vedic Science inspired the inquiry that led to this book, and whose Vedic Technologies expand human progress far beyond dependence on dangerous technologies—bringing life into full accord with natural law.

MAHARISHI MAHESH YOGI

Founder of the Transcendental Meditation and TM-Sidhi Program, Maharishi's Vedic Science, Maharishi International University, Maharishi Vedic Universities, Maharishi Ayur-Veda Universities, and Maharishi Universities of Management.

"This book reveals the poison to be purified and brings to light the nectar to revitalize life on earth. It is a textbook for everyone who is either a scientist or the beneficiary of scientific research, a book for everyone in the world today." —*Maharishi Mahesh Yogi*

TABLE OF CONTENTS

FOREWORD i

INTRODUCTION iv

CHAPTER 1
Overview: Health and the Environment in Jeopardy 1

The dangers of genetic engineering 1
- Health • Agriculture

Why is genetic engineering so dangerous? 6
- Unpredictable side effects • Irreversible mistakes

Genetic engineering: just speeding up a natural process? 9

Society's chronic failure to deal safely with new technologies 11

Examples from chemical and nuclear technologies 13

Averting the dangers of genetic engineering:
   Maharishi's Vedic Engineering 15

Moratorium Resolution 18

PART I: HEALTH

CHAPTER 2
**A Grim Prognosis: New Genetic Diseases
through Genetic Engineering** 21

Somatic cell gene therapy 22
- Technical limitations • Harmful effects of somatic
  cell gene manipulations • Other viable options

Germ-line gene therapy 33
- Harmful effects of germ-line genetic manipulations
- Human gene therapy not medically necessary

**Use of genetic engineering for diagnostics
   and for manufacture of drugs** 48
- Molecular diagnostics • Living factories—organisms
  genetically engineered to produce pharmaceuticals

The balance of benefit to harm 56

Summary 57

CHAPTER 3
Maharishi's Vedic Approach to Health: Freedom from Disease through Prevention 59
The case for prevention..................60
- The causes of almost all diseases are environmental and behavioral, not genetic • The solution: simple, natural prevention • A comprehensive prevention strategy is needed

Maharishi's Vedic Approach to Health................66
- A safe, effective approach for prevention and treatment of disease • Scientific evidence • Expanding research

The central principle..................72
- Nature's intelligence: the unified field of natural law
- Discovery of Veda in human physiology • Loss of connection with the unified field of natural law: the basic cause of disease

The central strategy................88
- Maharishi's Transcendental Meditation and TM-Sidhi program
- Restoring healthy functioning of the physiology: other modalities
- Maharishi Ayur-Veda Universities

Enlivening collective consciousness.................94
- The Maharishi Effect

Foundation for effective medical practice.................99

PART II: AGRICULTURE

CHAPTER 4
The Harvest of Agricultural Genetic Engineering: Genetic Pollution and Disruption of the Environment...............100

Extensive environmental release...................101
Lax regulations....................101
Hazards of environmental release of genetically engineered organisms for agricultural purposes.................102
Crops genetically engineered for herbicide resistance................103
- A strategy for increasing sales of agricultural chemicals
- Herbicide-resistant weeds • A clear case of "bait and switch"
- Making a profit wins out over humanitarian concerns

Genetically engineered bacteria can disrupt soil ecology 106
   • Alcohol-producing Klebsiella • Unpredictable
   effects of bacteria introduced into the soil
Genetically engineered livestock 108
   • Genetic pollution • Disruption of the food
   chain and the ecosystem
Genetically engineered squash 110
   • Destruction of biodiversity • More troublesome weeds
   • Generation of new viruses
Genetically engineered bovine growth hormone 113
   • Sick cows yield inferior milk • Synergy between
   over-optimistic scientists and over-eager businessmen
Partial solutions and novelty products 116

CHAPTER 5
Feeding the World without Genetic Engineering or Chemical Poisons—Maharishi's Vedic Approach to Sustainable Agriculture 119

Current challenges in agriculture 119
Sustainable agricultural strategies 121
   • Renewable fertilizers • Crop rotation and
   diversification • Soil conservation • Natural
   strategies for pest control • Research
Maharishi's Vedic approach solves agricultural problems 125
   • The farmer as manager • Improving managerial
   effectiveness • Wise farming • A more intelligent
   and creative social climate
Summary 132

INVITATION 134

REFERENCES 136

APPENDIX I 139
News conference, Washington, D.C., November 17, 1994
   • Selection of media coverage • News Release
   • Fact Sheet for Reporters • Questions from the press

APPENDIX II 169
Constitution of the Universe

FOREWORD

R. Keith Wallace

Ph.D., Executive Vice President
Maharishi International University

Few scientists today have taken an ethical stand on the potential dangers of genetic engineering, and none have ever before returned a federal grant in protest. On November 17, 1994, in Washington, D.C., Dr. John Fagan, Co-Director of the Doctoral Program in Physiology and Molecular and Cell Biology, and Professor of Molecular Biology at Maharishi International University, called for a 50-year moratorium on the environmental release of genetically engineered organisms. At the same time he formally announced the return of $614,000 in grant money awarded to him by the National Institutes of Health and withdrew grant proposals worth another $1.25 million.

Dr. Fagan is no ordinary scientist. He is a highly distinguished and serious researcher whose decision will have an impact on the entire scientific community. A number of top-level scientists openly supported Dr. Fagan's position and applauded him for his courageous act. After receiving his Ph.D. from Cornell University, Dr. Fagan worked for seven years at the National Cancer Institute as a postdoctoral fellow and then as a tenure-track scientist with his own research group. In 1984 he joined the faculty of Maharishi International University, an accredited university with 29 degree programs including six at the doctoral level. It was at Maharishi International University that Dr. Fagan received over $2.5 million in federal funding to conduct research on gene regulation and on the molecular mechanisms of carcinogenesis.

Dr. Fagan explains that his decision to return the money was the result of his growing concern about the applications of recombinant DNA research. Although these studies have provided a more fundamental understanding of the mechanisms of cellular and physiological functions at the level of DNA and of the roles that these mechanisms play in disease, Dr. Fagan feels strongly that too little research in this area has been completed to assure that either the current or potential future applications will be safe. In his own area of research to identify cancer susceptibility genes, he perceives that the only workable way this research could be used in medicine would be to alter the genetic material at the earliest stages of life. The consequences and potentially harmful side effects of such an action would be felt by all future generations. He feels that there are serious problems with the present system of regulation, and that it is a very real possibility that those genetically engineered plants and organisms which are now being released could cause widespread and deleterious effects in our environment.

The response to Dr. Fagan's ethical stand by national and international newspapers, television shows, and weekly magazines, as well as prestigious scientific journals, has been overwhelming.

Dr. Bevan Morris, President of Maharishi International University, fully supports Dr. Fagan's decision and made the following statement at the Washington press conference:

> "In October Dr. Fagan first mentioned to me his concerns about the dangers of genetic engineering and his growing concerns that the discoveries made in his own

research might indirectly contribute to those dangers. His sincerity and the depth of his thinking were striking. It made me and others at MIU reflect deeply on the responsibility of our university—and the responsibility of every university—to do research that is only supportive of life and health, now and for future generations.

"Our university has been proud of Dr. Fagan's research abilities and proud of the respect he has been accorded by his peers for his brilliant research into DNA, as reflected by the many research grants he has received. Now we are proud of him for giving up his grants and taking a strong stand on what he, as a scientist, believes to be right. Clearly he has shown he is motivated not by money or reputation, but by higher principles, and we at MIU feel proud of him for that.

"Dr. Fagan has always been given freedom to pursue his research directions at MIU, according to his scientific judgment and his conscience. This freedom is, of course, vital to the expression of his creativity as a scientist. Now, in this moment of decision, he is equally free to pursue the new direction he feels will be more productive for preventing and eliminating cancer.

"MIU will lose a great deal of money in the short term, but in my opinion nothing good will ever come to our university from money that is not going to unequivocally do good to the whole of life on earth. It is my hope that other researchers at other universities will be inspired by Dr. Fagan to reconsider their dependence on research grants from industry or government which force them to create technologies or drugs which may be profitable, but full of harmful side effects and long-range dangers to the life on our planet. Certainly, such a holistic view will be respected by future generations, while the short-term view of profits and technologies harmful to humanity will certainly leave future generations to curse the custodians of science of this generation."

INTRODUCTION

The first step toward writing this book began over 17 years ago, when I adopted recombinant DNA techniques in my own research and discovered their power firsthand. The recognition that this power could be used for harm as well as for good gave rise to concerns that have stayed with me over the years.

These techniques are not only used in research, but are also used in the applied science of genetic engineering. They enable the scientist, in essence, to carry out surgery on DNA. Just as a word processor can be used to cut out different pieces of text and splice them together to make new statements, a scientist can use recombinant DNA techniques to cut out different pieces of DNA and splice them together—recombine them—to make new statements in the genetic language. Using these techniques, the changes that can be made in the blueprint of any organism are, in principle, limited only by the imagination of the scientist or technologist using them. Thus, these techniques represent a powerful new set of tools that human beings can use to manipulate—for benefit or for harm—the structure and function of not only the human species, but of all other organisms on earth as well.

About two years ago my concerns escalated markedly when the popular press began to carry article after article that over-promoted genetic technologies, yet glossed over potential dangers.

INTRODUCTION

This newly aroused concern led me to re-examine my own research and to investigate how genetic engineering was being applied in medicine, agriculture, and industry.

Through this process it became clear that my research—with its long-term goal of identifying cancer susceptibility genes—could lead to medical applications that would create at least as many problems as they solved. (This is discussed in detail in Appendix 1.)

Likewise, it became clear that in medicine, agriculture, and industry, application of genetic engineering was moving forward at a dangerously rapid rate and that, despite some partial, limited benefits, products posing serious threats to the environment and to human health were quickly sliding through a lax and loop-hole-ridden regulatory system. It was clear that, if scientists and the public did not act quickly, these technologies would be widely implemented, without serious public debate. (These developments are discussed in detail in Chapters 1, 2, and 4.)

It was to stimulate such debate that I held a press conference in November 1994, where I returned a research grant to the National Institutes of Health, withdrew other grant proposals currently under review, warned the public about the impending dangers of genetic engineering, and encouraged scientists to take safer, more beneficial directions in their research. (The results of that press conference are reported in Appendix 1.)

The strong response of the scientific community, the news media, and the general public to this announcement has led to the organization of a *Global Alliance*. The immediate mission of this Global Alliance is to establish a worldwide, 50-year moratorium on germ-line genetic engineering in humans and on the environmental release of any genetically engineered organism. We invite everyone who is concerned about the safety and welfare of humanity to join in this alliance and carry this message to the world. (See Moratorium Resolution at the end of Chapter 1.)

In addition to recognizing the dangers of genetic engineering, two other insights motivated this action. The first was the very practical recognition that there were much more productive research directions that could be taken. As discussed in detail in Chapter 3, gene therapy, even at its best, can never address our most serious health problems. This is because the vast majority of diseases, including the big killers such as heart disease and cancer, are not genetic but have environmental and behavioral causes. For these, the most effective approach will not be high-tech gene therapy, but natural preventive measures that systematically attack the environmental and behavioral causes of disease. Recognizing this fact led to the decision, also discussed later in this book, to begin research into the prevention strategies of Maharishi's Vedic Approach to Health.

The second realization was more fundamental and relates to what is probably the greatest weakness of the modern scientific approach: namely, that the results of a scientist's labors can be used equally for good or for harm. Scientists have generally accepted this as an unavoidable feature of knowledge generated through the objective scientific approach. However, a scientist should not have to compromise in this way. The goal of a scientist's work should be to create knowledge that promotes life without shadowing it in any way, to create knowledge that is only for good.

Through experience at Maharishi International University, a strategy for addressing this weakness of the modern scientific approach has become apparent to me. For nearly 40 years, Maharishi Mahesh Yogi, the founder of our university and foremost scientist of the Vedic tradition of India, has worked to restore

INTRODUCTION

ancient Vedic knowledge and technologies. These technologies can be used as research tools to investigate the field of intelligence underlying and governing all of nature. The nature of the knowledge resulting from this investigation is such that it can be used only for good.

For those who are familiar only with the modern scientific paradigm, this is a surprising idea that requires some explanation. Scientists assume that the universe is orderly, that nature has structure and regularity. At the least, they assume comprehensibility in nature. They assume that there are natural laws, and then carry out objective research to uncover those laws, one by one.

In contrast, Maharishi's Vedic Science elucidates a deeper, more fundamental level of natural law. Each of the isolated laws of nature that modern science studies is a localized manifestation of this more fundamental level of nature's intelligence. While modern science studies these localized laws of nature in isolation, Maharishi's Vedic Science provides the tools to understand the integrated basis of all of those isolated laws, the most fundamental level of intelligence inherent in nature.

We often talk of basic research in the life sciences as research into the molecular mechanisms of those processes that are held in common by many different organisms. However, research that studies the intelligence underlying all of nature is an even deeper level of basic research. From the perspective of this knowledge, research into molecular mechanisms is not basic, but still deals with isolated and fragmented values.

Knowledge resulting from research into molecular mechanisms allows us to define procedures—technologies such as genetic engineering—that will have predictable effects. However, the scope of that predictability is limited. This is because any objective experiment studies only a piece of nature in isolation from the rest of the world. This isolation limits the scope of the knowledge elucidated by the experiment. Because of these limitations, the effects of applying this knowledge in the real world will be different from the effects of applying it in the laboratory to an isolated piece of nature. Application of this limited knowledge in the real world inevitably provides only partial solutions and creates problems as well as benefits. This is the source of the side effects and unexpected problems that arise when new technologies, such as genetic engineering, are applied.

For example, basic research on bacteria and other microorganisms has allowed scientists to design antibiotics that kill these microorganisms. However, these experiments did not allow scientists to predict or prevent the emergence of "superbugs" that are resistant to a wide range of antibiotics. As a result, dangerous infections—such as staph infections in hospitals—and communicable diseases, which we believed had been "conquered" many years ago by modern medicine, are reappearing in even more virulent forms.

Because Maharishi's Vedic Science studies the most basic level of nature's intelligence, a level that is common to all of nature, it generates knowledge that unifies our understanding of nature. This unified knowledge allows us to design technologies that spontaneously take everything in nature into account—nothing is left out of the equation. Technologies generated on the basis of this knowledge—Maharishi's Vedic Engineering—operate from a level common to all of life and thereby thoroughly solve the problem at hand and avoid harm to anyone or anything. These technologies give rise to globally beneficial results. The contrast between Maharishi's Vedic Engineering and genetic engineering is illustrated with respect to health in Figure 1.

INTRODUCTION

The purpose of this book is not only to warn of the dangers of genetic engineering but also to alert the public and the scientific community to the safer, more promising approach of Maharishi's Vedic Engineering. The time has come when we can choose a research direction that will contribute most powerfully to life on earth. Rather than continue on the dangerous level of fragmented knowledge of the physical structure of living things, it is possible at this time to explore the structure of intelligence within the physiology. It is time for scientists to begin investigating the basis of natural law rather than continuing to dissect little bits of the laws of nature.

We invite scientists everywhere to join with scientists at Maharishi Vedic Universities and Maharishi Ayur-Veda Universities worldwide, and at Maharishi International University, to carry forward this new and powerful research program.

Figures 1A. and 1B. Maharishi's Vedic Engineering and Genetic Engineering, Contrasting Approaches to Health (pages x and xi)
These diagrams schematically represent the hierarchical organization of the human physiology, showing how the unified field of nature's intelligence gives rise sequentially to the various levels of organization of the physiology. (A more detailed and complete representation is to be found on pages 75–77, that illustrates more thoroughly the various levels of organization and the connections between them.) Reading from the top of the chart, the human physiology as a whole is comprised of tissues and organs, which are in turn comprised of various cell types. Each cell consists of a vast array of biological molecules, represented by a diagram of the cellular metabolic pathways (intracellular metabolism), which convert nutrients into molecules needed for cellular growth and repair. Each dot in the diagram represents one of these molecules, and the lines connecting them represent the enzymes that interconvert these molecules. The structure of each of these enzymes is in turn specified by one of the thousands of genes within the cellular DNA. Seven of these genes are depicted, each with an arrow pointing to the enzyme whose structure it specifies. Complex as this diagram appears, it is actually a greatly simplified representation of physiological organization.

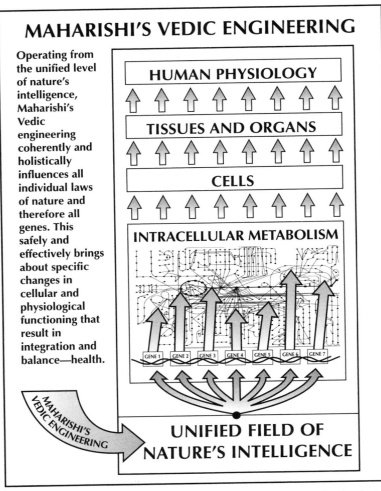

Figure 1A. Maharishi's Vedic Engineering is a systematic set of technologies based on Maharishi's Vedic Science. As indicated above by shaded arrows, these technologies bring about specific improvements in health by working from the level of the unified field of nature's intelligence to influence all genes, and all levels of organization of the cell and the physiology in an integrated and holistic manner. The links between this unified field and the physiology are discussed in more detail later. This approach integrates all aspects of the physiology so that they work together to support health. At the same time, because this approach takes all components of the physiology into account, no harmful side effects occur.

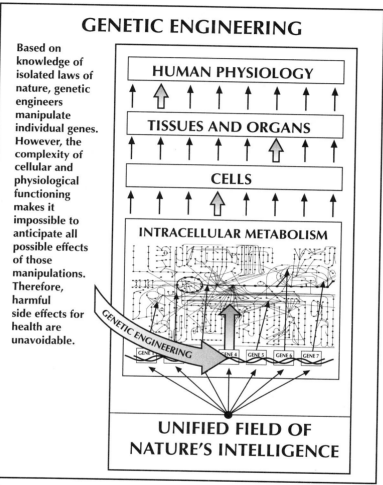

Figure 1B. Gene therapy, a branch of genetic engineering, attempts to treat disease by altering specific genes. The rationale of this approach is that all of the molecules that comprise each cell are either proteins themselves, or are synthesized by proteins (enzymes). In turn, all proteins are encoded by genes. Therefore, it should be possible to modify the functioning of cells and of the physiology as a whole—restore health—by altering the genes for certain proteins.

However, this approach yields only short-term, partial success, because health is the result of integrated, balanced functioning of all aspects of the cell and of the body. Balance cannot be restored by manipulating isolated components of the cell. For instance, altering Gene 4 will change the amount or activity

(continued on page xii)

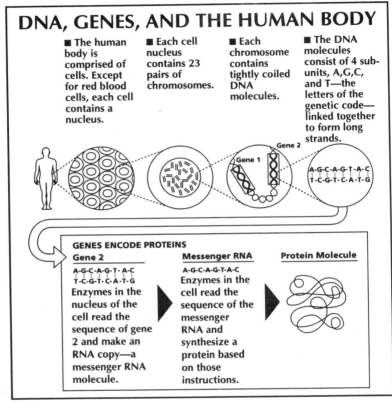

Figure 2. DNA, Genes, and the Human Body
Every component of the human body is made of proteins or of other molecules that are synthesized by proteins (enzymes). This diagram shows how these building blocks of the body are made. (See page 4 for more details.)

Figure 1B. *(continued from page xi)*
of the corresponding enzyme. As indicated in Figure 1B. by shaded arrows, this will result in only isolated changes in intracellular metabolism and in the functioning of the cells, tissues, and organs of the body. These isolated changes may improve one area, yet create new imbalances in other areas. Therefore, this approach provides only partial and temporary solutions to health problems. Moreover, genetic manipulations will inevitably lead to unanticipated harmful side effects: Because the cell is an integrated system, changing one gene inevitably influences other cellular components, and because this system is highly complex, it is impossible to predict the ultimate effects on health.

CHAPTER 1

OVERVIEW: HEALTH AND THE ENVIRONMENT IN JEOPARDY

This book has two purposes. The first is to sound a warning concerning the dangers of genetic engineering to human health, agriculture, and the environment. The second is to point out that the technologies of Maharishi's Vedic Science—referred to in this book as Maharishi's Vedic Engineering—offer far safer, more effective, and more economical approaches. These can solve, today, the problems that genetic engineers unrealistically claim they can solve after many more years of research and development. This first chapter provides an overview, presenting in brief the central issues discussed more thoroughly in later chapters.

The dangers of genetic engineering

At present, there are powerful economic and political forces driving the widespread implementation of genetic technologies.

The race is in full swing to capitalize on these technologies and to recoup the billions of dollars that have already been invested in research and development. In the rush to make the most of this opportunity, government leaders, industrialists, and scientists are underestimating and, in many cases, disregarding the dangers of genetic engineering.

To a large extent, the public and government leaders are not even aware of the gravity of the situation. To create and expand the market for genetic technologies, there has been, and continues to be, a vigorous effort to shape public opinion in favor of these technologies. The popular press has been flooded with articles that oversell genetic engineering. These articles tend to gloss over serious risks and play on the noble sentiments of the population, promising that genetic engineering will:

- Solve all our health problems—For instance, one of the leading gene therapists has stated that someday in the not-too-distant future, physicians will send patients home cured of most diseases simply by injecting them with a "snippet" of DNA (1).
- Conquer hunger—Proponents claim that genetically engineered crops will increase agricultural productivity.
- Improve the environment—Proponents claim that genetic technologies will reduce the use of agricultural chemicals and offer improved methods of dealing with environmental pollution.

Based on these claims this approach sounds like the best thing that has ever happened to humanity, but when examined more closely, we find abundant evidence that each of these claims is extravagantly over-blown. Each ignores fundamental limitations of genetic technologies. Each ignores the serious dangers of these technologies. These are briefly described in the following paragraphs and discussed in detail in later chapters.

Health

Gene therapy is the use of genetic engineering in attempts to correct genetic defects that are responsible for specific diseases. Today over a hundred clinical trials are in progress to develop gene therapies for heritable diseases, cancer, and AIDS. In addition to over-promotion in the media, a more troubling trend is beginning to emerge: namely, that many leaders in the biomedical sciences are now proposing that we should pursue not only somatic gene therapy but also germ-line gene therapy.

Germ-line gene therapy would alter the genes of the germ, or reproductive cells of an individual. Based on what is already being done in other species, the most straightforward approach would be to inject new DNA—a new gene—into the nucleus of a fertilized human egg (early embryo). Subsequently, this new gene would be incorporated into the embryo's own DNA. When the embryo developed to maturity, each cell of that person, including their reproductive cells, would carry the new DNA. Because this procedure would alter the DNA of reproductive, or germ cells, the person would pass that alteration on to all subsequent generations. Another approach to germ-line manipulations would be to alter the genes of eggs or sperm before fertilization occurs.

Somatic cell gene therapy involves the introduction of new genes into any cell of the body other than reproductive cells. For instance, to treat cystic fibrosis, scientists are attempting to alter the genes of lung cells; to treat adenosine deaminase deficiency (ADA)—the bubble boy disease—they are altering the genes of white blood cells.

Gene therapy leads to the following dangers, which, along with other health risks, are discussed in detail in Chapter 2:

Some Definitions

DNA—DNA, deoxyribonucleic acid, is a long, double-stranded molecule whose function is to store biological information (Figure 2, p. xii). DNA is composed of four sub-units, called adenosine, thymidine, guanosine, and cytidine, abbreviated A, T, G, and C. Many of these sub-units join together like the links of a chain to make up the strands of a DNA molecule. Genetic information is stored in the sequence of these sub-units. They can be considered the four "letters" of the genetic alphabet, or genetic code. The sequence in which these sub-units appear in the DNA molecule carries information in exactly the same way that the sequence of letters on this page carries information.

Gene—a statement in the genetic language that specifies the structure of a protein. Thus every structure and every component of an organism is either made of proteins or made of other molecules that were themselves synthesized by proteins (enzymes). Genes carry, in either explicit or implicit form, essentially all of the information specifying the structures of that organism. (There are other inputs that influence the development of an organism and the evolution of a species.)

Genome—the collection of genes specifying all of the structures and functions of a particular organism. In higher organisms, the genome is located in the nucleus of every cell.

Genetic engineering—the use of recombinant DNA techniques to alter the genes of an organism for medical, agricultural, industrial, or other purposes. Genetic technologies are the products of such alterations.

Recombinant DNA techniques—a set of powerful enzymatic, microbiological, and chemical methods that enable the scientist, in essence, to carry out surgery on DNA. Just as a word processor can be used to cut out different pieces of text and splice them together to make new statements, a scientist can use recombinant DNA techniques to cut out different pieces of DNA and splice them together—recombine them—to make new statements in the genetic language. Using these techniques, the changes that can be made in the blueprint of any organism are, in principle, limited only by the imagination of the scientist or technologist using them. Thus, these techniques represent a powerful new set of tools that human beings can use to manipulate—for benefit or harm—the structure and function of not only the human species but also all other organisms on earth.

- Genetic manipulations could inadvertently mutate normal genes. This could disrupt essential cellular or physiological functions, causing new genetic diseases or cancer.

- Many proteins are known to have more than one function. Thus, when genetic engineers alter a gene for the purpose of changing one function of a protein, they may inadvertently interfere with a second function, thereby impairing essential cellular or physiological processes, leading to a new genetic disease or cancer.

- Germ-line genetic manipulations alter the individual's germ cells, or reproductive cells (egg or sperm). Therefore, harmful side effects will be passed on to all subsequent generations. Attempts to improve health could thus create new inherited diseases, birth defects, and cancer susceptibility.

- Widespread application of germ-line genetic manipulations would result in an ever-expanding problem of genetic pollution in which interactions between genetically engineered genes and other genes, either natural or genetically engineered, would lead to unanticipated, harmful side effects.

- Misuse of genetic technologies for cosmetic or social purposes, as distinct from medical purposes, would result in abuses of human rights and generate new medical and social problems. Also, if germ-line manipulations were involved, this would greatly accelerate the problem of genetic pollution.

Agriculture

To date, only a few genetically engineered organisms have been released into the environment on a large scale. However, hundreds more have been developed and are poised for clearance by regulatory agencies in the U.S. and many other countries, for use in applications that could result in either their purposeful or accidental release into the environment. This is a grave threat to the integrity, balance, and future evolution of life on earth. The following are some of the most significant risks:

- disruption of the ecosystem, either locally or globally, leading to loss of biodiversity, disruption of the food chain, and destruction of centers of biodiversity for important food crops;
- creation of new plant diseases, new pests, and new weed varieties that are resistant to existing chemicals;
- genetic pollution—accidental introduction of defective genes into the gene pool, which could weaken the vigor and fitness of a species;
- disruption of soil ecology and reduction of soil fertility;
- increased use of toxic, carcinogenic, and mutagenic agricultural chemicals, leading to water pollution and, in turn, to increased incidence of cancer, birth defects, and other illnesses.

Why is genetic engineering so dangerous?

Of the technologies now in use, genetic engineering is especially dangerous because many of the most common applications of this new technology threaten to generate unexpected, harmful side effects that cannot be reversed or corrected, but will afflict all future generations. The side effects caused by genetic manipulations are not just long-term. They are permanent.

Unpredictable side effects

It is true that genetic engineers are now capable of altering DNA with reasonable precision. However, these alterations are defined only in physical and biochemical terms; genetic engineers cannot fully and reliably predict the biological effects of these alterations. They cannot adequately predict how these manipulations will influence cellular functioning, the physiology and behavior of the organism as a whole, and the ecosystem into which that genetically engineered organism will be introduced.

It is impossible to confidently predict the effects of genetic manipulations because of the complexity and interconnectedness of living systems. Whether we examine the simplest single-celled microorganism, or a human being, or the global ecosystem, we find a huge number of complex components. These take part in extremely intricate, coordinated interactions, all as part of one, vast, integrated, unified phenomenon—life.

For instance, within the human body we find organ systems, which consist of different groups of organs. These are made up of tissues, which are in turn composed of cells. These levels of organization are interrelated and interconnected in both form and function. In order to carry out its function within the whole, each component is completely interdependent with the others. For example, in order for the lungs to fulfill their function, they depend on the heart and the circulatory system, and, likewise, those organs depend on the lungs.

Looking deeper, within any one of the trillions of cells that make up our body, there is another vast world of complex subcellular structures, organelles, molecular networks, and metabolic pathways, each composed of a variety of biomolecules. All work together in an integrated, interdependent manner.

On the global scale, we find a similar situation. The countless forms of life that populate the earth are not isolated, autonomous entities. They comprise one huge, enormously complex and dynamic fabric of life. These organisms participate in the continual, interactive exchange of intelligence, energy, and materials, enabling each to obtain what it needs in order to grow, reproduce, and evolve. Thus their lives are interconnected in infinitely many ways.

For instance, a moth may be absolutely dependent on a particular flowering plant as a food source, while that plant may be equally dependent on the moth to pollinate its flowers. Many such interactions comprise the microfabric from which the wider web of life is constructed. These interrelationships have been fine-tuned through millions of years of co-evolution, and this process is continuing.

We see that all living systems are highly complex and at the same time unified and integrated. The degree of complexity is so great that genetic engineers cannot take all components of the system into account. Yet, when they change any single component, they inevitably influence the system as a whole. In such a situation, surprises are inevitable and experience shows that many of those surprises will not be pleasant ones.

Irreversible mistakes

Unexpected, negative side effects are typical, not just of genetic engineering, but of all technologies emerging from the objective scientific approach. The difference between genetic engineering and other technologies is the *duration* of negative effects. Although many of the earlier technological impacts, such as nuclear and chemical pollution, seem long-term from the perspective of the human time scale, on geological and ecological time scales they are relatively temporary. When we terminate the activities causing chemical or nuclear pollution, mechanisms come into play that, in time, restore the ecosystem. For example, when we stop using harmful chlorofluorocarbons, the damage done to the ozone layer by these compounds will be repaired by natural photochemical reactions in the upper atmosphere, which generate new ozone.

In contrast, germ-line genetic manipulations alter the reproductive cells of the organism. As a consequence, those alterations will be passed on to all subsequent generations. Mistakes made during those manipulations and harmful side effects that they cause will not dissipate with time, but will be perpetuated.

Once released, genetically engineered organisms cannot be recalled or eradicated from the environment. They will, in many cases, propagate and persist for long periods of time and move, either passively or actively, to other locations. This will vastly expand the range of environmental and climatic conditions, and the range of other organisms, with which a genetically engineered organism will interact. It will also greatly extend the time frame within which those interactions can occur. Because of the interconnectedness of all life forms, these interactions will in time propagate throughout the global ecosystem. The complexities of these potential interactions and the extended time frame in which they can occur create virtually infinite possibilities for unpredicted consequences. Thus, no matter how small a particular genetic alteration may seem, no matter how insignificant to the ecosystem an organism may appear, that alteration will not only irrevocably alter the course of evolution of that species, but will also, over the long term, distort the whole fabric of life and flow of evolution.

Genetic engineering: just speeding up a natural process?

Proponents' favorite response to concerns about genetic engineering is that genetic change is a natural process and that genetic engineers are doing nothing that Mother Nature is not already doing herself; they are just speeding up the process and giving it a certain direction. In the same breath proponents say that these genetic manipulations are certainly no different from the classical plant and animal breeding strategies that humanity has pursued for thousands of years.

Both of these arguments miss—or evade—the point. It is true that genetic change is natural. However, natural genetic change is extremely restricted in its scope and rate. Mutations and rearrangements of the DNA can occur and genetic information can be transferred from closely related species under special circumstances. However, these are incremental changes. They occur in minute steps that are strictly circumscribed by a host of biological and physical constraints. In contrast, genetic engineers are far less restricted. In principle and in practice, they are limited only by their own imaginations. They can splice together genetic information from any two or more organisms on the planet and can introduce that DNA into virtually any organism that they choose. When a genetic engineer splices a bacterial gene into the genome of a squash plant, he has accomplished something that would require hundreds of thousands, if not millions, of years of natural genetic transformations. Practically speaking, such transformations would never take place in nature. The fact that genetic engineering alters the genetic material is not the uniquely objectionable feature of this approach. What raises concern is the unbridled nature of these changes, along with their unprecedented scope and their potential for creating unpredictable side effects.

It is also true that genetic engineering and traditional breeding are similar in that they both involve human manipulation or interference with other living things. But there the similarity ends. Traditional breeding could be likened to economic cooperation between two countries, while genetic manipulation is more like forceful military take-over. Species boundaries and other biological and physical constraints place very stringent limitations on what can be done with traditional breeding. The breeder must work within those limitations, working with, cooperating with, nature. Consequently, there are many things that the breeder may desire to accomplish that are simply not possible within those limitations. In contrast, current methods of gene manipulation cut through all of those boundaries, and future technological developments will make it even easier for genetic engineers to accomplish whatever their imaginations dictate. This lack of boundaries is the source of the greater dangers of genetic engineering.

Yes, genetic change is natural, but the magnitude and scope of the changes that can be brought about through genetic engineering are much greater than those that occur naturally or through traditional plant and animal breeding. As a consequence, the effects of these manipulations on the organism and subsequently on the environment are much less predictable. It is almost certain that if environmental release of genetically altered agricultural varieties becomes widespread, or if human germ-line genetic manipulations become common, unanticipated damaging side effects will occur—if not sooner, then later. Furthermore, once altered genes enter the gene pool, they cannot be recalled; the problems that they create will be irreversible.

Society's chronic failure to deal safely with new technologies

Scientists and leaders have proposed that we address the potential dangers of genetic technologies in the same way that we have dealt with the hazards of chemical and nuclear technologies in the past. They propose to make laws to regulate the use of these technologies.

But what has been the result of using this approach with chemical and nuclear technologies? Past experience shows that the ability of scientists and industrialists to envision new applications has always outstripped safety considerations. The race to capitalize on discoveries in the nuclear and chemical sciences spurred rapid and widespread implementation. Only later did we recognize the serious, often deadly, side effects of those applications.

Technologies develop along the following lines:
1. Basic research leads to new conceptual breakthroughs.
2. Scientists point out potential applications of this new knowledge, claiming that they promise miraculous benefits.
3. Business and industry hurriedly capitalize on this new knowledge, implementing applications on a large, commercial scale before scientific evidence regarding possible dangers is available.
4. Once applied on a large scale, serious side effects become apparent.
5. Society retrenches, often at great cost, and demands research concerning the dangers of the technology.
6. Research results are then used to guide safer application of the technology.

This pattern has been repeated over and over again with chemical and nuclear technologies. The basic problem is that an initial discovery itself often points to apparently attractive commercial applications, while much more research is often required before sufficient information is available to appreciate the potential dangers of those applications for health and the environment. This problem is compounded by the fact that research and development of commercial applications offer economic incentives, while research on safety does not. In fact, strategic ignorance—that is, the convenient lack of evidence concerning dangers—allows industry to claim, "We do not know of dangers associated with this technology, therefore there is no reason not to implement it."

Examples from chemical and nuclear technologies

There are numerous cases of serious side effects resulting from applications that were attractive because of their commercial potential, but were damaging to human health or the environment. If more research had been done before large-scale commercial implementation, many lives would have been saved and much suffering avoided.

For example, in the 1950s X-ray machines were installed in shoe stores to assist in fitting shoes, until it was discovered that chronic X-ray exposure causes cancer and other problems. Certainly this application was an effective tool for selling shoes. However, it offered no substantive benefit to the individual or society that justified the suffering that resulted from its use. Thus, this technology could have been implemented only in ignorance of the health dangers of X-ray exposure. Similar problems arose from using X-ray machines as security devices. X-rays were also used to treat a wide variety of diseases, including tuberculosis and enlarged thymus, only to find that this treatment increased the incidence of several forms of cancer.

Another example is the use of radium dials on gauges and watches. Those who used such dials suffered somewhat increased cancer risk due to radiation exposure. However, the people who suffered most were those who applied the radioactive paint, often by hand. Chronic absorption through the skin or accidental ingestion of radium-laced paint resulted in excruciatingly painful and deadly bone cancer.

Nuclear power is another technology which was implemented without sufficient safety assessment. Today the citizens of many nations have come to the conclusion that the risks of this technology are too great to justify its benefits. It has been realized that even if we incorporate all possible safeguards into nuclear power plants and even if we structure the most elaborate and conservative safety procedures for the operation of those plants, there is a small but real risk that a serious accident will occur, resulting in the release of radioactive materials into the environment. Probability theory tells us that if many power plants operate over an extended period of time, it is highly likely that even very low probability risks will become realities. The Three Mile Island and Chernobyl accidents confirm this conclusion. In addition, nuclear power plants generate large amounts of radioactive wastes that cannot be disposed of, but must be stored, creating huge long-term risks. The public has realized these risks and is saying with much greater frequency, "Not in my backyard!"

There are also abundant examples of chemical technologies causing harm to human health or the environment when they are implemented before sufficient research results are available.

- When chlorofluorocarbons first began to be used we did not know enough about their chemistry to anticipate that they would, with the help of ultraviolet light, react with ozone in the atmosphere, thereby contributing significantly to thinning of the ozone layer.

- When lead additives were first introduced to improve the properties of gasoline, who would have predicted that their combustion products would react with other atmospheric pollutants to greatly exacerbate the smog problem?

- In the 1940s, when DDT was sprayed directly on children as an insecticide, as well as in abundant doses onto our food, there was no evidence that this compound and many other halogenated hydrocarbons have hormone-mimetic properties, that they accumulate in the food chain, and that they could seriously compromise the health of humans and wildlife, even bringing certain species of birds to the brink of extinction. Even today this class of compounds continues to affect human health adversely and may be leading to reduced reproductive capacity.

- When it was first marketed in the 1950s, who would have anticipated that thalidomide, prescribed as an antiemetic for pregnant women, would have disastrous effects on fetal development? This is one of hundreds of examples of unexpected, negative side effects of pharmaceuticals.

In light of the negative effects of implementing these chemical technologies, the phrase "better living through chemistry," which was once taken at face value, has now taken on strongly ironic connotations.

Averting the dangers of genetic engineering: Maharishi's Vedic Engineering

From the examples above, we see that with nuclear and chemical technologies, society has been and is even today trapped in a destructive pattern that links technological development to harmful side effects. The question that we must ask is this: If we take this same approach with genetic technologies, is there any reason to expect a different outcome? The answer is, of course, "NO."

We have gotten away with this approach with chemical and nuclear technologies because they do not affect life on earth at as deep a level as genetic technologies, and because their effects are more short-term. Genetic engineering alters an extremely fundamental level of life, and once a genetically modified organism is released into the environment, those alterations will be passed on to all generations: they will be irreversible. If we follow the same path with genetic engineering that we have followed with earlier technologies, irreparable damage will be done.

Is there an alternative?

One approach would be simply to move forward much more slowly and methodically, with greater care. Toward this end, we propose a 50-year moratorium on the large-scale implementation of any genetic technology that might, either intentionally or accidentally, lead to the introduction of genetically engineered genes into the global gene pool of any species—humans, animals, plants, or microorganisms. This moratorium will provide a cooling-off period during which research can be carried out to assess the safety and appropriateness of genetic technologies, and during which our leaders and the public can more systematically evaluate all options before adopting a new technology whose effects are irreversible.

This approach should and must be taken, but something more is needed. Experience shows that the pressures driving commercialization will always oppose this strategy. They will inevitably accelerate the rate of implementation and push the commercialization of applications that will be profitable in the short term, but harmful to health or the environment in the long run.

If we limit our efforts to a moratorium and to research on the safety of genetic technologies, serious impacts on health and the environment will be inevitable. Moreover, as discussed in the introduction, the fundamental problem will still remain: Knowledge generated through the current scientific paradigm unavoidably creates side effects. This same paradigm creates a style of thinking in which society is willing to accept the damage caused by side effects as the unavoidable price that we must pay for "progress." In the final analysis, using this approach by itself will only perpetuate the problem.

In addition to a moratorium and research on safety, we need to come to grips with the fact that harmful, unpredictable side effects are unavoidable with technologies based on the modern scientific paradigm. Therefore it is necessary to look to technologies derived from other systems of knowledge—systems that do not have this limitation.

As discussed in the introduction to this book, Maharishi's Vedic Science is such a system. It offers scientifically validated technologies, collectively termed Maharishi's Vedic Engineering (discussed in more depth in Chapters 3 and 5). These can safely and effectively address, today, the problems in health and agriculture that genetic engineers unrealistically claim that they can solve with years more of research and development.

In light of this, the draft of a Moratorium Resolution, presented below, not only calls for a halt to certain applications of genetic engineering, and for research on safety; it also calls for the implementation of these safer, more effective Vedic technologies. The likely outcome of this moratorium will be that we will discover that it is unnecessary to subject society to the potential dangers of genetic technologies.

CALL FOR A MORATORIUM ON:
- **Germ-Line Genetic Engineering in Humans**
- **Environmental Release of Genetically Altered Organisms**

Resolution—We call for a 50-year moratorium on: (1) germ-line genetic manipulations in humans; and (2) any application in which genetically engineered organisms may be released into the environment either accidentally or intentionally.

The purpose of this moratorium will be to allow time for adequate research on the safety of genetic technologies, and to allow time to implement other safer, more effective approaches in medicine, agriculture, and industry.

We hold that intelligent application of currently available knowledge can solve even the most difficult challenges confronting humanity today, thereby making implementation of genetic technologies unnecessary.

Dangers of genetic technologies—Enthusiasts claim that genetic engineering can solve virtually every problem facing the world today. However, closer scrutiny reveals that these claims are naive and over-optimistic. In fact, these claims are dangerous since scientific evidence indicates that wide application of genetic engineering: (1) will be harmful to human health; (2) will threaten the global ecosystem; and (3) could be socially destructive.

These dangers arise because germ-line genetic engineering alters the genes of the germ, or reproductive, cells of an organism. As a consequence, altered genes will be passed on to all subsequent generations. Errors and side effects caused by genetic manipulations will enter the gene pool for that species and will be perpetuated *ad infinitum*, creating new genetic diseases. Moreover, through unanticipated interactions with other organisms, genetically engineered organisms can have potentially large, unpredictable effects on the ecosystem. This amounts to genetic pollution. Once in the gene pool, genetic errors are irreversible and uncontrollable and therefore will have much more harmful, pervasive, and longer-lasting effects than chemical or even nuclear pollution.

Safer, more effective solutions—In addition to its potential for creating extremely long-lived harmful side effects, genetic engineering can, at best, offer only partial, temporary solutions to our health and agricultural problems. It is unnecessary to subject humanity to these risks since safer and more effective approaches are available today.

Solutions in health and medicine—The most serious and pressing problems of contemporary medicine are complex diseases such as cardiovascular disease and cancer. These and the vast majority of other diseases will never yield to genetic engineering because they are due primarily to environmental and behavioral causes, not genetic defects. For these diseases, the most effective approach will not be high-tech gene therapy, but preventive measures that systematically attack the environmental and behavioral causes of disease.

Maharishi's Vedic Approach to Health offers a coherent, comprehensive, and integrated strategy for prevention—something that our therapeutically-oriented medical system lacks. Prevention is the central focus of this approach. It includes holistic, integrated approaches in therapy, as well. This time-tested, scientifically validated natural health care system is capable of eliminating disease at its source, rather than merely managing later stages of disease and treating symptoms. Twenty years of research has established that this system can successfully address even the most serious and pervasive health problems, including cardiovascular disease, cancer, drug abuse, stress syndromes, and many chronic diseases (*Scientific Research on the Transcendental Meditation and TM-Sidhi Program: Collected Papers, Volumes 1-6*—over 500 studies and 4000 pages).

Solutions in agriculture—In agriculture, the innovations proposed by genetic engineers are dangerous and superfluous. These innovations are dangerous because the environmental release of genetically engineered organisms and the flow of altered genes into native populations pose serious risks to both the local and global ecosystems. They are superfluous because effective, natural, sustainable agricultural methods are already available and ready for wide-scale implementation today. These methods can immediately meet or exceed the levels of productivity and food quality that genetic engineering naively aspires to provide in the distant future.

We propose to develop optimal strategies for large-scale implementation of an integrated, coherent system of sustainable agriculture. Maharishi's Vedic Science provides principles that can serve as the organizing center of this system. This system will be capable of immediately meeting the goals of increased farm production and enhanced food quality and nutrition that are the stated, but inaccessible, goals of agricultural genetic engineering. In contrast to current farming practices that are dependent on chemicals and deplete the soil, this system will foster self-sufficiency in the farmer, and preserve and enhance agricultural resources and the environment.

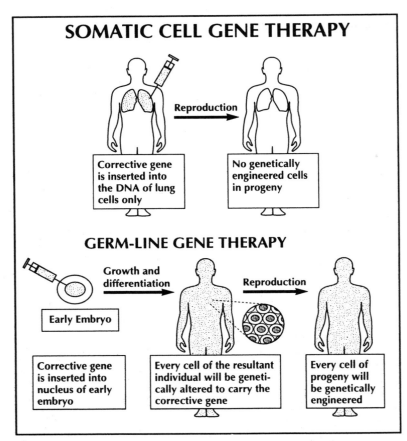

Figure 3. Somatic Cell Gene Therapy and Germ-Line Gene Therapy
Somatic gene therapy alters the genes of cells other than reproductive cells, for instance the lung cells of an individual with cystic fibrosis. Because reproductive cells are avoided, the genetic alteration will not be passed on to subsequent generations. Germ-line gene therapy alters reproductive cells, and therefore the genetic alteration will be transmitted to all future generations.

PART I: HEALTH

CHAPTER 2

A GRIM PROGNOSIS: NEW GENETIC DISEASES THROUGH GENETIC ENGINEERING

There are two approaches for the direct use of genetic engineering in medicine: somatic cell gene therapy and germ-line gene therapy. There are, of course, other *indirect* applications, such as the use of genetically engineered organisms to produce drugs and vaccines, and the use of recombinant DNA techniques to develop molecular diagnostics.

These applications all have potential benefits that have been widely publicized and that will be touched on later in this chapter. These applications also pose potential dangers, which have been publicized much less broadly. The objective of this chapter is to delineate these dangers in more detail and to consider the appropriateness of implementing these technologies, in light of both the risks and benefits. We will focus primarily on gene therapy because this application is the most hazardous.

The strategy of gene therapy is to correct defective genes that cause or contribute substantially to certain diseases. As illustrated in Figure 3, germ-line gene therapy alters the reproductive or germ cells of the organism, while somatic cell gene therapy modifies cells other than reproductive cells, such as liver, brain, muscle, and skin cells. Thus, somatic manipulations will not be transmitted to subsequent generations, while germ-line genetic modifications will. On one hand this limits the potential improvements resulting from somatic cell gene therapy to the single individual who undergoes treatment. On the other hand, mistakes or harmful side effects resulting from somatic cell gene manipulations will harm only one individual, whereas mistakes resulting from germ-line manipulations mistakes will enter the human gene pool and jeopardize the health of all future generations. The following sections discuss these approaches in more detail.

Somatic Cell Gene Therapy

Over 100 somatic cell gene therapy programs have been initiated since 1989 in the United States. Research is moving forward with similar intensity in other areas of the world as well. Current efforts focus on developing somatic cell gene therapies to combat cancer and AIDS and to correct inherited diseases that can be traced to a defect in a single gene. These are called monogenic diseases.

Technical limitations

Many enthusiasts claim that somatic cell gene therapy will eventually become the treatment of choice, not only for monogenic inherited diseases, but also for a wide range of other diseases. Some workers in the field even claim that this approach has promise for treating infectious diseases (2). At present, however, researchers are encountering huge obstacles in successfully treating even monogenic inherited diseases; much greater difficulties will be encountered with other diseases. Despite the millions of dollars and the thousands of researcher-years that have been invested in somatic gene therapy to date, not a single clear-cut clinical success has been reported. Clinical trials have provided useful research data, but they have not been successful therapeutically. For instance, adenosine deaminase (ADA) gene therapy patients have always been maintained on enzyme replacement therapy, in addition to gene therapy. The performance of these patients has been neither better nor worse than that of patients maintained on enzyme therapy alone. Thus it is not accurate to claim that this treatment has been clinically successful.

Although somatic gene therapies may be developed for some diseases, the technical limitations discussed below make it unlikely that this will ever become a general approach for treating even monogenic diseases, not to mention other diseases—especially those with more complex causes.

Currently, most somatic cell gene therapy trials make use of an *ex vivo* strategy (3). As shown in Figure 4, this strategy involves: (1) removing cells from the patient's body; (2) introducing into those cells a correct copy of the gene for the missing enzyme or protein; and (3) reimplanting those cells in the patient's body. Those cells then produce the enzyme or protein that was previously lacking. Researchers have already used this approach with lymphocytes and more recently with bone-marrow-derived stem cells isolated from the circulation of several children with ADA deficiency (2, 3, 4) which if untreated causes severe immunodeficiency and

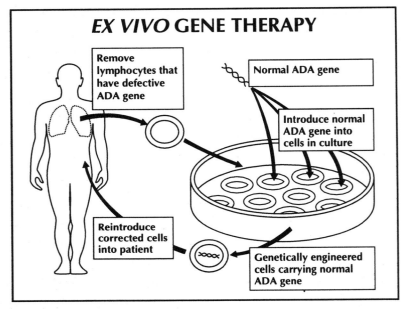

Figure 4. *Ex Vivo* Gene Therapy
Ex vivo therapy is currently the predominant approach in somatic gene therapy, for instance, in treating for ADA deficiency. In this disease, children who are unable to make the enzyme adenosine deaminase (ADA) die at an early age because their immune systems cannot fight infections. This is because, without the ADA enzyme, their white blood cells—lymphocytes—cannot survive long enough to fight the disease. Genetic engineers are attempting to correct this problem by removing lymphocytes from the patient, introducing a healthy ADA gene into those cells, and then introducing the cells carrying the healthy ADA gene back into the patient's blood.

---

leads to early death.

The *ex vivo* strategy is feasible for only a few of the many monogenic diseases: those that involve either a cell type that can be isolated and then reimplanted, or those that are due to the systemic lack of some protein that could be supplied by any accessible organ of the body via the circulation. Use of somatic cell gene therapy for other diseases will require the delivery of new, correct genes to specific cells that are part of a specific tissue; or it will require the delivery of new genetic material to all, or a large percentage of, the cells of the body. The new genetic information must not only find its way into one or more specific cell populations, but must also be incorporated into the cell in a manner that allows it to be functional so that it can program those cells to express the protein or enzyme which was previously lacking. In addition, the expression of that gene must be properly controlled so that the cell synthesizes the proteins at the appropriate times and in the appropriate amounts.

The technical challenges in developing the sophisticated gene delivery and gene expression systems that are required to make somatic gene therapy workable are formidable. Initial work with the most tractable systems is now in progress (for instance, delivery of the cystic fibrosis gene to the patient's lung cells via inhaled aerosols), but the development of generally applicable and highly efficient, accurate, and specific gene delivery and expression systems has hardly begun.

To meet these stringent requirements for gene delivery and gene expression, genetic engineers will need to have very specific and detailed basic knowledge of the biochemistry, cell biology, and molecular biology of the particular cell type to which they wish to deliver the corrective gene. Knowledge of general principles will not be sufficient. In practice, the requirement for extensive fundamental knowledge—specific to a particular disease and to the cell types involved in that disease—means that it is unlikely that we will be able to develop a generalized strategy for somatic gene therapy. Moreover, the depth of knowledge required will greatly limit the rate at which gene therapy protocols customized to specific diseases can be created and may make it impractical to consider developing gene therapies for some diseases.

In summary, huge investments of resources and time will be required to transform the current speculative vision of somatic gene therapy into a practical clinical reality.

Harmful effects of somatic cell gene manipulations

Mutations

Gene manipulations designed to correct genetic defects in somatic cells carry the risk of causing mutations to cellular DNA. This risk derives primarily from the imprecision of the currently available methods for inserting genes into cellular DNA. With the methods currently used in somatic gene therapy, the therapist cannot insert the new genetic information into a defined site within the cellular DNA. Instead, it is introduced at a random location, as illustrated in Figure 5. This means that every gene therapy procedure is potentially mutagenic. If the new genetic material is accidentally inserted into the middle of some other gene, that gene will be inactivated. If the new genetic material is inserted in close proximity to another gene, it may alter DNA sequences that control the expression of that gene, thereby increasing or decreasing the level of expression of that gene. Future methodological refinements are likely to provide the genetic engineer with better control, but the risk will always be substantial that genetic manipulations will cause unintended mutations.

In most cases such mutational events will not cause serious problems. In a typical somatic cell gene therapy treatment, new DNA will be introduced into a billion or more cells, but it will be

NEW GENETIC DISEASES THROUGH GENETIC ENGINEERING

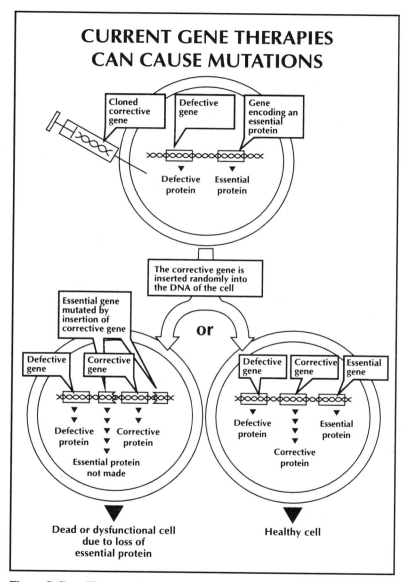

Figure 5. Gene Therapy Can Mutate Genes Essential for Health and Survival
Currently, gene therapists can only insert genes at random into the cellular DNA. Thus, therapy carries the risk of mutating genes that are important to the health or survival of the cell.

incorporated into a different site in the DNA of each cell. In a few of those cells, the therapeutic gene will be inserted into the middle of an important cellular gene, destroying its function and causing that cell to die or lose its ability to carry out its normal role in the physiology. However, experience indicates that in most cases insertion does not noticeably damage cellular functioning.

The one case in which problems could and will arise is when mutations are induced in genes involved in cancer. It is known that mutations in some genes (proto-oncogenes and tumor suppressor genes) can convert normal cells into cancer cells. Since a single cancer cell can lead to tumor formation, a somatic gene therapy treatment can cause cancer, if it happens to induce a carcinogenic mutation in even one cell. The risk of carcinogenic mutations is quite significant, because a single somatic gene therapy treatment will alter billions of cells, any one of which could sustain a mutation that could turn it into a cancer cell. In every gene therapy procedure there are billions of opportunities to generate a carcinogenic mutation. In addition, many somatic cell gene therapies will have effects lasting only a few months, and therefore will have to be repeated periodically as the altered cells are replaced by new cells. This has already been found to be the case with ADA gene therapy. This means that the patient will be subjected again and again to a mutagenic treatment that will affect huge cell populations. The probability that these repeated treatments will trigger a carcinogenic mutation over a period of a few years is reasonably high.

Side effects

Somatic cell gene therapy also carries with it the risk typical of any pharmacological intervention, namely that of unanticipated negative side effects. In addition to the intended therapeutic effect, the newly incorporated genetic information could have unanticipated harmful effects on cellular or physiological functioning. For instance, somatic cell gene manipulations will often result in cellular production of a protein that the patient's immune system has never encountered before. The patient may mount an immune reaction to this new protein, which may neutralize that protein's therapeutic capacity. Furthermore, the immune system may attack the cells producing that protein with damaging effects on health. A second example of possible unanticipated effects is in the use of somatic cell gene manipulations to sensitize the immune system to cancer cells. This procedure is designed to stimulate the immune system to destroy cancer cells, but it may also result in auto immune reactions—the sensitized immune system may actually attack the normal cells from which the cancer arose.

Because gene therapy acts at such a fundamental level of physiological organization—the level of the DNA itself—the range of possible side effects is quite broad, and our ability to anticipate and avoid problems is very limited. Thus, in the process of correcting one problem, somatic cell gene therapy may decrease the well-being of the patient in other, unanticipated ways. Since a single treatment may strongly influence the physiology for several months, and in some cases may be permanent, these side effects could have a substantial negative impact on the patient.

Public opinion

Probably the greatest risk that accompanies somatic cell gene therapy—although it is itself little more dangerous than many other experimental drugs under investigation today—is that it can serve as a catalyst to promote acceptance of germ-line gene therapy, which *does* carry great risks. This can and is already happening via two mechanisms.

First, somatic cell gene therapy can be used as a tool to mold public opinion in favor of gene therapy in general, and eventually in favor of germ-line therapies. A few partial successes with somatic cell gene therapies, along with sustained supportive publicity on gene therapy in general, will serve to popularize this approach and create a climate in which the public is more receptive to all forms of gene therapy, including germ-line approaches. This effect is already at work, as attested by a recent survey in Great Britain (5). There, it was found that the portion of the population willing to use gene therapy to control violent behavior, alcoholism, and other problems increased more than three-fold in a two-year period. The survey authors attribute the change to the wide publicity received in the intervening period by research on gene therapy for cystic fibrosis (CF). Public opinion has moved in this direction as a result of exposure in the popular press that has over-emphasized and exaggerated the potential benefits of gene therapy, while glossing over potential dangers, disadvantages, and limitations. In place of such coverage, what is really needed is well-informed, open discussion of gene therapy in which all segments of society participate fully in critically considering all sides of the issue.

The second mechanism by which somatic gene therapy can catalyze a transition to germ-line therapy is as follows. Government and the private sector are investing quite substantially in somatic cell gene therapy, yet the gene delivery and other technical problems with somatic cell gene therapy, discussed above, make this approach a technical blind alley. Somatic cell gene therapy is equivalent to the steam automobile of the turn of the century—clumsy, expensive, and restricted in applicability. Furthermore, the investment in time, money, and manpower required to move

NEW GENETIC DISEASES THROUGH GENETIC ENGINEERING

beyond these limitations is extremely large. In contrast, the technical challenges with germ-line therapy are more modest. Investment in somatic cell gene therapy will create tremendous pressure —when the fundamental limitations and inadequacies of somatic cell gene therapy are finally acknowledged—to recoup those investments by moving into germ-line strategies. Germ-line techniques are being actively developed using other species. In the next few years, breakthroughs will inevitably occur that make germ-line manipulations feasible in humans. At that point, if the climate of public opinion is receptive to gene therapy, a transition to the germ-line approach will be almost inevitable.

Other viable options

We have established two points. First, somatic gene therapy carries substantial risks, and second, success in developing effective somatic gene therapies is far from certain because of formidable technological obstacles.

These points bring up an important question: Do the potential benefits of somatic cell gene therapy justify the costs embodied in these risks and uncertainties? This question must be considered separately for two classes of diseases—monogenic inherited disorders, such as ADA and cystic fibrosis; and other diseases such as cancer, AIDS, and cardiovascular disease.

The majority of gene therapy clinical trials currently in progress are for cancer, AIDS, and cardiovascular disease. However, the answer to the above question in relation to these diseases is unequivocally "No." As we will discuss in detail in the later part of this chapter and in Chapter 3, there are other far more practical and cost-effective approaches to addressing these problems. The primary reason that gene therapists are currently focusing on these diseases is economic. Funds are abundant to support research on cancer, AIDS, and heart disease, while support is much more restricted for research on monogenic inherited diseases. Therefore, in order to finance their work, researchers whose primary interest is the development of gene therapy methods have promoted the tenuous hypothesis that gene therapy can be useful in addressing these diseases.

For monogenic inherited disorders, the question posed above is more complex. Many of these diseases are either deadly or seriously debilitating. At the same time, effective therapeutic approaches for these diseases are not currently available. Since they are fundamentally genetic diseases, it has been relatively easy to argue that somatic cell gene therapy is a valid strategy for addressing them. However, when one considers the many uncertainties associated with the development of successful gene therapies, and that this approach carries significant risk of side effects, it becomes clear that it is important to actively and exhaustively explore other approaches to addressing these diseases, as well.

As will be discussed later in this chapter, there is, in fact, great promise for the discovery of non-genetic approaches to treating these diseases. Such approaches will be only palliative, since they cannot correct underlying genetic lesions. However, this becomes less of a disadvantage when one realizes that many gene therapy strategies will also be only palliative. For instance, the current gene therapy strategy for cystic fibrosis will correct the CF defect only temporarily and only in lung cells. It will be necessary to repeat therapy frequently—possibly monthly. Furthermore, this therapy will correct the defect only in the patient's lungs, leaving untouched other critical sites where CF is known to cause dysfunction, such as in the kidneys. Thus the patient will not be "cured" by gene therapy.

In addition, non-genetic therapies have two important advantages. First, they are less invasive. This means that risks to the patient will be reduced. Second, they are much more "low-tech" than gene therapy. This means that they will be less expensive, and therefore accessible to a much broader spectrum of the population. Since some of the most prevalent genetic disorders, such as sickle cell disease and thalassemia, are endemic in less developed regions of the world, cost-effectiveness is a critical issue.

In light of these considerations, it is essential to commit substantial research resources to the development of approaches other than somatic gene therapies for treating monogenic diseases.

Germ-line gene therapy

Already five European countries—Austria, France, Germany, Norway, and Switzerland—have laws on the books that ban germ-line genetic engineering in humans (6). But in the U.S., no such law exists. The Recombinant DNA Advisory Committee of the National Institutes of Health administers regulations that forbid researchers at institutions receiving government support from carrying out human germ-line manipulations. All others, including commercial laboratories, are encouraged to follow these regulations but are under no legal requirement to do so. Regulations in many countries are even more lax, or non-existent.

There is a wide range of opinion among scientists, ethicists, and legal experts regarding human germ-line genetic engineering. As familiarity with the concept has grown, there has been a transition from the opinion that human germ-line genetic engineering simply should not be considered to a stance that makes consideration contingent primarily on the availability of workable technology. This stance is typified by the following statement in a report from a committee of the British Royal Society of Medicine, whose responsibility is to consider the ethics of gene therapy: "All members of the committee were agreed that germ-line therapy should not be attempted at the present time. This agreement was not based on some moral or ethical precept, but solely on the safety factor: there were too many unknowns in the possible outcome (7)."

Some scientists (including W.F. Anderson, one of the pioneers in the gene therapy field) have for years been promoting a much more aggressive viewpoint (8). They hold that, when methods are better developed, germ-line genetic engineering should be pursued in humans. This view ignores the serious human rights and social issues related to germ-line manipulations, not to mention the danger that errors or side effects of such manipulations could generate new heritable diseases. Yet, a growing number of scientists and ethicists are favoring this view. This became apparent in public discussions that were triggered recently by the announcement of advances in reproductive biology that will make it possible to manipulate the genetic content of sperm. Regarding this development, Ralph Brinster, whose research triggered this debate said, "the genie is out of the bottle (9)." The supporters of this trend include some of the most respected and influential members of the scientific community. For instance, Daniel Koshland—editor of the leading journal *Science,* member of the U.S. National Academy of Sciences, and past president of the American Society of Biochemistry and Molecular Biology—has spoken out repeatedly in favor of using genetic approaches to deal with a wide range of social as well as medical problems (10,11). Thus there is, today, growing pressure in support of the development and use of germ-line genetic technologies.

Even in light of these developments, it might be argued that the following discussion is irrelevant, because a workable technology for carrying out germ-line manipulations in humans is not available today: "Why waste time talking about germ-line therapy when we couldn't accomplish it even if we wanted to?" Quite to the contrary, it is critical that we examine these possibilities now if we are to shape future directions for research and development. It is especially important that we send signals to the commercial sector that discourage heavy investments in human genetic technologies. The drive to recoup such investments would create tremendous pressures to carry forward with commercialization, despite safety considerations.

The dangers of human germ-line genetic engineering fall into two categories. The first is unanticipated harmful effects to individual and public health. The second is harm that could come from diverting this technology from medical applications to cosmetic uses, and to social or political purposes. These are discussed in more detail below.

Harmful effects of germ-line genetic manipulations on health

Germ-line gene manipulations carry four distinct risks: (1) harmful mutations—simple errors that occur during the process of carrying out genetic manipulations that destroy or alter genetic information in harmful ways; (2) unanticipated effects of genetic alterations on the health or welfare of the individual; (3) genetic pollution due to unanticipated interactions of the altered gene with other genes in the human gene pool; and (4) interference with the natural course of human evolution.

Harmful mutations caused by genetic manipulations

As discussed earlier, germ-line or somatic genetic manipulations can inadvertently induce mutations in important genes—because of the imprecision of genetic engineering techniques—thereby harming the health of the individual. However, two factors greatly amplify the dangers with germ-line manipulations. First, since the manipulation is done very early in development of the embryo, essentially every cell in the organism will be genetically modified in exactly the same way. If the genetic manipulation procedure happens to inactivate an essential gene, that gene will be inactive in every cell of the patient's body. This will have a much greater impact on the physiology as a whole than would be the case with somatic gene therapy. With that approach, genetic material is inserted at a different site in each cell, and therefore only a small fraction of those cells are likely to be genetically harmed by the treatment.

The second factor, which amplifies the gravity of errors, is that germ-line manipulations alter the reproductive cells. Thus, errors caused by germ-line genetic manipulations will not only be transmitted to every cell of the individual's body, but will also be transmitted to that individual's children and to each subsequent generation. In effect, germ-line manipulations carry the risk of creating new heritable genetic diseases, birth defects, and cancer.

Unanticipated side effects of germ-line genetic manipulations

Even if new genetic material is inserted into the genome without causing mutations, there is still the danger that, once inserted and expressed, this new genetic material could have unanticipated effects on cellular and physiological functioning. These, too, could translate into new inherited genetic diseases or new heritable susceptibilities for birth defects or cancer.

The human physiology is simply too complex to allow us to predict how newly introduced genetic material will interact with other genes and influence cellular and physiological functioning. Thus, unpredictable negative side effects are almost inevitable.

Another source of unanticipated side effects comes from the fact that genes, and the proteins that they encode, often serve multiple functions. We may identify a gene on the basis of one function and be completely unaware that it has other essential functions. This happens because we know only a tiny portion of what there is to be known about the human physiology and about the genes that serve as its blueprints. If we modify a gene to correct a defect in a known function, we may inadvertently interfere with another essential but unknown function of that gene, thereby doing more harm than good.

The gene affected could encode an essential enzyme whose loss would result in a debilitating or fatal metabolic disease. On the other hand, the function affected could contribute to some valuable trait, such as intelligence, artistic creativity, mathematical or musical ability, etc. Loss of such a function might diminish those traits. Consequently, gene therapy could lead to trade-offs that would be quite unfortunate for both the individual and society. For instance, the fact that many of our most famous writers and artists have also been alcoholic may indicate that some gene that influences susceptibility to alcoholism may also influence artistic creativity. If this is the case, which is pure speculation at the moment, when scientists identify a genetic defect related to alcoholism and use that information to correct susceptibility to that problem, the resultant child, and its progeny, may have lost unique qualities of artistic creativity. In "purifying" this family line of a trait that predisposes to alcoholism, genetic engineers may have deprived that family and humanity as a whole of something of great value. Yet this is the kind of social application of gene therapy that Koshland and others advocate (see earlier reference). Again, we remind the reader that this example is speculation. However, the fact remains that we really cannot isolate and correct for one function without affecting other functions. Basically, we just do not know enough.

This reveals a fundamental defect in the current tendency to think about disease in terms of genetics. From the current perspective, alcoholism can be viewed as a disease caused by mutations in one or more specific genes. On the other hand, the view can be taken that the gene is normal but that current social, environmental, and behavioral patterns create a certain constellation of stresses that induces the disease called alcoholism in individuals carrying that gene. From this perspective, it is short-sighted, and dangerous for humanity, to attempt to correct stress-induced diseases by altering genes that happen to influence susceptibility to that stress. It is better to correct the environmental causes of those diseases.

The following example puts this point into sharper focus. Polycyclic aromatic hydrocarbons are one of the largest and most prevalent classes of environmental pollutants. They are generated by internal combustion engines, by manufacturing processes, by waste incineration, and by the smoking of tobacco. Evidence is mounting that there are specific genes that confer on some individuals the ability to protect themselves more effectively from the harmful effects of polycyclic aromatic hydrocarbons.

The appropriate approach to dealing with environmental pollution from such compounds is to clean up the environment and reduce the amount of the compounds released into the air and water. However, another approach would be to identify the gene that confers resistance and then use germ-line genetic manipulations to transfer that gene to everyone in the next generation, thereby giving them the ability to survive in an environment burdened with higher levels of polycyclic aromatic hydrocarbon pollution. This approach is quite analogous to correcting the alcoholism gene instead of addressing the environmental and life-style stresses that induce this problem.

Many medical conditions have a genetic component—not only alcoholism, but also cardiovascular disease, cancer, depression, schizophrenia and others. However, to claim that gene therapy is the solution to these problems is at best naive, wishful thinking; more likely it is self-serving promotion by those who wish to justify research and development of human gene therapy. Such ideas are expressed in almost every popular article or professional review of gene therapy. These are destructive ideas for two reasons. First, they encourage us to postpone coming to grips with the environmental and behavioral stressors that are the real causes of these diseases. Second, if the gene therapy strategy is implemented in medicine, humanity will be much the poorer due to genetic trade-offs like the hypothetical trade-off between alcoholism and artistic creativity.

Pollution of the human gene pool

If germ-line genetic engineering were to become common, harmful mutations due to these manipulations would accumulate in the human gene pool, as would genetic modifications that had negative side effects. This amounts to genetic pollution.

Once the door is open to human germ-line genetic engineering, there will be a continuing and widening stream of altered genes entering the human gene pool. With this the potential for unpredicted interactions between different genetically engineered genes—and between genetically engineered genes and natural ones—will increase exponentially.

Some argue that because the human gene pool is vast, the new genetic defects that might be created as by-products of germ-line genetic engineering would be swallowed up in that vastness and have a negligible effect on humanity as a whole (12). This is, in effect, saying that the gene pool is so huge that one more genetic disease will not matter. It is surprising to hear such arguments from individuals who advocate spending millions of dollars on and devoting thousands of researcher-years to attempting to eradicate currently existing genetic diseases. It just does not seem productive to replace one genetic disease with another. In considering how much genetic pollution is acceptable, the safest approach is not to open the door to this possibility in the first place.

Interference with the natural course of human evolution

Through the spontaneous process of natural selection, nature has been fine-tuning, revising, and polishing the genetic code script of humanity and of all other species for an extremely long period of time. To propose that we should revise this text reveals the naive and arrogant assumption that we have grasped that text fully; or that we have, at least, understood it sufficiently so that we can see how it can be improved. We are not in that position today. As pointed out above, we do not even have good criteria for distinguishing true genetic defects from positive traits that happen to be vulnerable to contemporary environmental, social, and behavioral stressors. Using the tools of science, we have at best gained small glimpses into the organizing principles and the structure of this vast text. We have much yet to learn before we are ready to improve on nature by altering the human genome.

There is another level of consideration: human evolution is an ongoing, sequentially unfolding process. Germ-line genetic manipulations will intercede in this process in a very artificial and unnatural way. If germ-line manipulations become common, alterations will be carried out that would never have occurred naturally. When viewed from cellular, physiological, and biological time scales, many of these alterations will appear to be minor and innocuous. However, on the evolutionary time scale, they may have profound—and completely unanticipated—consequences. They may trigger sequences of events that would never have happened as part of the natural process of evolution, thereby irrevocably changing the direction of human evolution. From our current level of ignorance—we do not even know the sequence of five percent of the human genome at this time—we have no way of predicting whether those changes would be for better or for worse. Knowing this, how can we in good conscience meddle with the genome?

Dangers of the cosmetic, social, and political applications of gene manipulations

The second class of dangers associated with manipulating the human genome would emerge if this technology were used to ostensibly improve or enhance human functioning, instead of being used to correct debilitating genetic defects. Either somatic-cell or germ-line methods could be used for this purpose. However, the most significant dangers would accompany the use of germ-line manipulations. Either private citizens or social-political organizations might be motivated to use this approach to alter the characteristics of the next generation in ways that they consider desirable, resulting in the hazards outlined below.

If procedures for human germ-line genetic manipulations were developed for medical purposes, two pressures would drive their use for non-medical applications. First, the market for gene therapy is much smaller than the market for genetic enhancement. Thus, industry will be strongly motivated to expand to this larger market to increase profits. No longer would application of this technology be limited to those few who have rare genetic diseases. Instead, every potential parent would become a potential customer. As will be described below, this pressure has already subverted one product of genetic engineering—human growth hormone—to cosmetic uses, with potentially harmful effects on children's health. The second pressure that would drive cosmetic applications is the simple desire of parents for the best offspring possible. Commercial interests could exploit this desire to build a huge market for genetic improvements. Creating such a market would generate a situation in which millions of germ-line manipulations would be done. This would profoundly accelerate the process of genetic pollution, leading, in a generation or two, to a wide range of unanticipated, damaging side effects.

It is true that parents have a right to make use of all possible means for ensuring that they will have healthy children. However, there is a vast difference between access to adequate prenatal care and access to methods for the genetic reprogramming of one's offspring. The right of parents to fulfill their desires to control the characteristics of their child must be balanced by the individual rights of the child and by the collective rights of all future generations. The child will have to live throughout life with the consequences, both negative and positive, of the parents' decision to

NEW GENETIC DISEASES THROUGH GENETIC ENGINEERING

manipulate their child's genome. Few of us would happily concede to our parents the life-long right to control something as superficial as the clothes that we wear or our hair style. How much more uncomfortable would it be to have another individual's tastes imposed on us regarding something as intimate and inescapable as our physiological or psychological makeup. On the broader scale, any germ-line genetic manipulation of humans will inevitably enter and thereby alter the human gene pool. This gene pool is the collective property of all humanity for all time to come. Does any one individual have the right to tinker with it?

To place powerful genetic technologies at the service of a particular political or social belief system opens up another whole range of possible misuses. These could cause extensive, possibly irreversible, damage to society, not to mention accelerating further the process of genetic pollution described above. Such applications bring with them the potential for huge abuses of human rights. During World War II, a socially and politically motivated genetic agenda led to millions of deaths. Such agendas are unfortunately not a thing of the past; statutory eugenics programs are ongoing in at least two countries today.

Such policies give rise to two problems. First, they violate the fundamental human right to control one's own reproduction. The second problem has to do with the criteria for determining what is "inferior." In instituting a eugenics program, one generation is in essence imposing its values and vision of human existence, not just on the next generation, but on all generations to come. No generation has the right to do this.

Human gene therapy is not medically necessary

A final argument can be made against human germ-line genetic engineering: This procedure is not needed medically. The medical objectives that this technology aspires to achieve, but cannot achieve any time in the near future, can already be met through other medical approaches that are available today. Moreover, the currently available methods are far safer than germ-line genetic manipulations. The following sections document these points.

High-tech genetic manipulations are not necessarily the best possible medical treatment

One argument often raised in favor of human germ-line genetic engineering is that physicians are obligated to provide the best available treatment for their patients. But what is meant by the best available treatment? Does that necessarily mean the newest, most high-tech methods available? Instead of assuming that the newest is best, we need to scientifically evaluate all possible methods to determine—based on objective evidence—which ones are, in fact, best. Such an analysis immediately leads to the conclusion that gene therapy is not the best approach for the vast majority of diseases, because most diseases—98–99%—are primarily due to environmental and behavioral causes, not genetic defects. As will be discussed in the next chapter, the best medical approach for these diseases is prevention.

Non-genetic treatments for genetic diseases

But what about the 1–2% of diseases that are genetic? Some researchers hold that for these, gene therapy is the long-term answer. However, even for many of these diseases, effective alternative approaches exist now or could be developed. For instance, controlling diet during childhood prevents the negative effects of the genetic disease phenylketonuria. In this disease, toxic levels of phenylalanine accumulate due to lack of one of the enzymes important in breaking down this amino acid. By reducing phenylalanine intake, these toxic effects can be prevented. Likewise, Wilson's disease, which is due to a defect in a protein that helps the body remove excess copper, and which leads to lethal hepatitis and neurological degeneration, can be corrected completely by treatment with drugs that assist in the excretion of copper. Why risk the creation of genetic diseases that would harm many future generations, when simple dietary or pharmaceutical measures will satisfactorily solve these particular problems?

Even molecular medicine provides evidence to support the possibility of effective non-genetic treatments for genetic diseases. For instance, the severity of sickle-cell anemia varies strikingly from patient to patient, even though all patients have the same genetic defect—a mutation in the beta hemoglobin gene. How could the same genetic defect result in such wide variations in the severity of this disease? Although other possibilities exist, one explanation is that the severity of symptoms is influenced by the environmental and behavioral conditions experienced by different patients. If this is the case, these variations serve as a clue suggesting that modifying those environmental or behavioral influences (*i.e.*, through diet or pharmaceuticals) could be the basis for effective, non-genetic approaches to treating this fundamentally genetic disease. This line of reasoning has recently been validated by development of a drug, hydroxyurea, that effectively and safely alleviates the symptoms of sickle cell disease. This drug seems to work by activating another hemoglobin gene—gamma-globin—which takes over the function of the defective beta-globin gene.

Such variations in severity of symptoms are not the exception but the rule with genetic diseases, thus offering hope for the success of this strategy. Another clue supporting this strategy is the observation that many genetic diseases, particularly those classified as autosomal dominant diseases, manifest their symptoms only later in life and do so at variable times. If it is possible to avoid symptoms for many years, why not for one's whole life? We just need to learn how to manipulate environmental factors to further delay or even prevent onset of symptoms. These clues, along with current successes in treating some genetic diseases—such as phenylketonuria, Wilson's disease, and sickle cell anemia—clearly leave open the possibility that non-genetic therapies can be developed for many, if not all, genetic diseases.

Much more research is needed in this area. The biomedical sciences would yield practical health knowledge much more productively if even a small fraction of the money now allocated for research on gene therapy were redirected to support research on non-genetic approaches to treat genetic diseases and research on preventive approaches to deal with the majority of diseases.

There is even a safer, high-tech alternative

Modern, high-tech medicine provides two other alternatives that obviate the need for human germ-line genetic engineering. The first of these alternatives, which is now beginning to be used in some countries, is the use of in vitro fertilization coupled with in vitro diagnosis. Embryos from a couple who carry a genetic disease can be fertilized in vitro, and tested in vitro by sensitive genetic diagnostic methods to identify an embryo that is free of the genetic defect. After implantation, this embryo will develop into a healthy child. The second approach, which is in wide use today, is prenatal diagnosis followed by termination of pregnancy if genetic defects are found. Virtually any couple who found germ-line genetic manipulations acceptable would also find these approaches acceptable.

In pointing out these options, we are not advocating them. There are highly complex ethical, social, and human rights issues associated with these approaches. We are simply pointing out that germ-line genetic engineering shares all of these ethical, social, and human rights issues, yet has an additional serious problem. It carries the risk of creating health-damaging genetic errors and side effects that can be passed on to all future generations. This health and safety hazard makes human germ-line genetic engineering even less acceptable than the other alternatives mentioned above.

In light of the fact that proponents of germ-line genetic engineering are aware of these two alternatives, one can only conclude that they are purposefully misrepresenting the facts to the public when they state that germ-line manipulations would fill a unique and practically useful purpose in the physician's black bag.

Human germ-line genetic manipulations, no valid justification

The above points establish that arguments justifying human germ-line genetic engineering on the basis of medical need are spurious. If this technology is not medically essential, then why are some scientific leaders advocating its development and use?

Do these scientists possess political or social motives? Are they motivated by hope of financial gains from commercializing this technology for cosmetic purposes? Although there may be an occasional genetic engineer motivated by the former two, and a few more motivated by the latter, in general another explanation is most likely. That is, scientists have failed to rigorously and realistically evaluate the potential contributions of gene therapy. Therefore, they have not recognized that this approach is not likely to contribute substantially to improving human health. Moreover, they have failed to appreciate fully the potential side effects and misuses that this technology will inevitably engender, and the destructive effect that it will have on human life. It is essential that scientists act responsibly at this time, and weigh more critically the potential benefits and the dangerous consequences of their research.

Use of genetic engineering for diagnostics and for the manufacture of drugs

Genetic engineering is also used in medicine in the manufacture of drugs, hormones, vaccines, and dietary supplements, and in the development of sensitive diagnostic tests for genetic defects. These applications do not involve altering the human genome, nor do they involve the release of genetically engineered organisms into the environment. Therefore, the risks associated with these applications are much smaller than those associated with gene therapy or agricultural uses of genetic engineering.

Molecular diagnostics

Almost every week we open the newspapers to find another article announcing the discovery of the gene related to another genetic disease—muscular dystrophy, cystic fibrosis, hemophilia, Huntington's disease—the list goes on and on. Characterization of these genes gives scientists the basic information that they need to develop very specific, precise, and sensitive diagnostic tests for these diseases. With the discovery of genes that contribute to susceptibility to cancer, cardiovascular disease, and other diseases, genetic diagnostic tests for these diseases can also be developed.

NEW GENETIC DISEASES THROUGH GENETIC ENGINEERING

For instance, the discovery of the BRCA1 gene makes it possible to identify those women predisposed to breast cancer due to mutations in that gene.

In some cases—those where diagnosis can be followed up with positive, effective therapy—these new diagnostic procedures can be of value. For instance, the damage which phenylketonuria causes to the nervous system can be completely prevented if children carrying this defect can be identified early in life and placed on a modified diet. However, there are negative impacts of these procedures as well. One problem is that diagnostics are outstripping therapy. There are diagnostic tests today for many diseases for which modern medicine lacks adequate therapy. Such tests have no medical utility and may negatively affect the patient's psychological health and well-being.

An even more critical problem is that the information from these powerful diagnostics can be misused by insurance companies, employers, and others. Within a few years diagnostic tests will be available for literally hundreds of genetic diseases and for genes that influence susceptibility to hundreds of other diseases, such as cancer, heart disease, and schizophrenia. Such a battery of tests would be of great value to insurance companies and employers, because they could use it to eliminate potential policy-holders and employees who are likely to have major medical problems.

From the perspective of the individual, however, this use of genetic information constitutes a serious violation of personal privacy. In most areas of the world, there are no laws regulating the use of such information, and there are already many cases in which such information has been misused to the detriment of the individual.

This is recognized as a serious problem, and there is at present a concerted effort to solve it. It is likely that satisfactory solutions will be developed and implemented in the next few years, probably in the form of laws regulating the use of the information derived from these diagnostics. In fact, such laws have already been passed in some states of the U.S., though the means to enforce them are not established.

Although this problem is significant, it is clearly of much less weight than those discussed above, which are related to the use of gene manipulations for non-medical purposes and to alter the human germ-line. Unfortunately, at this time, organized efforts to explore the ethics and social impacts of genetic technologies are focused almost exclusively on the genetic information issue. For instance, the ELSI (Ethical, Legal and Social Issues) program of the U.S. Human Genome Project offers support for research regarding issues related to genetic information, but has not currently funded research regarding questions surrounding germ-line genetic therapy. Is the genetic information issue being used as a smoke screen to distract from these other, even more substantial issues?

Living factories—organisms genetically engineered to produce pharmaceuticals

Many strategies have been devised to engineer organisms to produce medically useful substances. Most of these strategies make use of genetically engineered microorganisms. For instance, as illustrated in Figure 6, these methods have been used to insert the gene for human insulin into the DNA of bacteria. This manipulation converts these bacteria into microscopic factories that produce human insulin. When grown in large numbers in a fermenter, they can be used to produce large amounts of insulin that can be sold to treat diabetes.

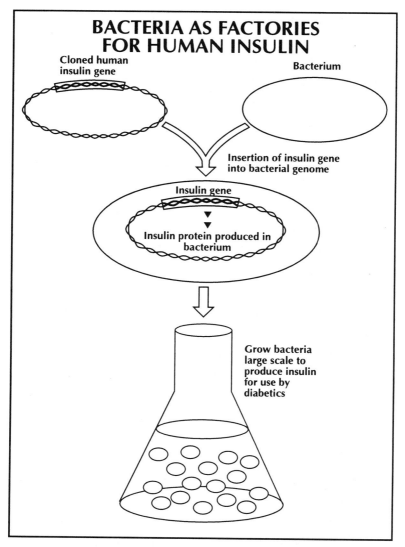

Figure 6. Bacteria as Factories for Human Insulin
The insulin gene has now been isolated, or cloned, from humans, and inserted into bacteria. When these bacteria are grown on a large scale, they produce large amounts of insulin that can be purified and used to treat diabetes.

These applications do not alter the human genome and do not involve the purposeful release of genetically engineered organisms into the environment. Furthermore, the risk of accidental release is small, because the organisms used are purposely weakened so that they cannot survive in the wild. Thus, the direct risks associated with these genetically engineered processes are small. The actual risks associated with these applications are related, not to the manufacturing process, but to the properties of the products. That is, the risks are related to the possible negative side effects of the drugs, hormones, and vaccines produced using these processes. Thus, the risks are similar to those of ordinary pharmaceuticals.

Modern medicine evaluates drugs on the basis of the ratio of beneficial to harmful effects. Because the risks associated with these genetically engineered pharmaceuticals are relatively low compared to gene therapy, the ratio of beneficial to harmful effects is actually favorable for some products.

For example, in the treatment of diabetes, human insulin produced in genetically engineered bacteria is superior to insulin extracted from porcine and equine tissue. It has fewer side effects, because it is free of the contaminants—which can cause immune reactions—present in the earlier product. Since the genetically engineered insulin is produced in bacteria, it is also more acceptable to vegetarians and to others who observe cultural or religious restrictions in diet, and is more economical to produce.

Likewise, recombinant vaccines are clearly an advance over traditional vaccines. They are far safer and more uniform in their characteristics, although they can have side effects. Recombinant tissue plasminogen activator, an enzyme that breaks down blood clots in arteries, has been found useful in treating stroke and heart attack victims. Another successful product is granulocyte-macrophage colony stimulating factor (GMCSF), which helps restore the function of the immune system in chemotherapy patients by stimulating growth of lymphocyte stem cells. Tissue plasminogen activator and GMCSF do, of course, have side effects, but they have significantly improved survival of stroke, heart attack, and chemotherapy patients, thus offering a favorable benefit-to-harm ratio.

Other applications are relatively neutral—no great contribution to humanity, some commercial advantages, but no serious side effects. For still other applications the ratio of beneficial to harmful effects is clearly negative. The following two examples are useful because they illustrate the dangers that can arise and also provide insights which are relevant to the more serious hazards associated with gene therapy.

Genetic manipulations to customize a bacterium for efficient tryptophan production

Several companies use bacterial fermentation to produce tryptophan which is sold as a human nutritional supplement. To increase the efficiency of tryptophan production, one company, Showa Denko, carried out extensive genetic modifications of the bacterium used in this process. The tryptophan produced by this genetically engineered organism was tested according to U.S. Food and Drug Administration standards, and was cleared for commercial production and sale as a food supplement for humans.

Not satisfied with the efficiency of tryptophan production, this company carried out further genetic modifications of this organism. These manipulations were successful in greatly increasing output. However, Showa Denko decided not to carry out the costly series of safety tests that were done with the earlier genetically engineered microorganism. The company assumed, based on the earlier tests, that the tryptophan produced using the new bacterium was safe and immediately began selling it as a nutritional supplement.

Over the next two years 37 people died, another 1500 were partially paralyzed, and 5000 more were temporarily disabled by a syndrome that was finally linked to this company's tryptophan (13). It was contaminated with one or more toxic compounds chemically related to tryptophan.

Most likely, the genetic manipulations had increased tryptophan production so greatly that the concentrations of tryptophan within the bacteria reached very high levels. As a result, the tryptophan and its precursors began to react chemically producing unexpected toxic compounds. To date this company has paid over $1 billion in damages, and litigation is still in progress.

There is still controversy regarding the cause of this toxicity, since, in addition to using the new genetically engineered bacterium, this company had also begun to cut corners in the procedure for purifying their tryptophan. To protect the good name of genetic engineering, the industry has blamed the toxicity on these procedural changes, and the company has destroyed the bacterium, so that further research cannot be done. However, scientists who have followed this incident favor the explanation in which genetic manipulations caused the bacteria to generate toxic tryptophan derivatives.

The question that this example raises is this: If it is impossible to predict the consequences of altering the genes of a simple, single-celled organism, how can genetic engineers say with confidence that manipulations of the far more complex human genome will improve health and not create new diseases?

Recombinant human growth hormone stimulates growth in children

Genentech, one of the leaders in the biotech industry, has genetically engineered bacteria to produce human growth hormone in large quantities. The appropriate medical use of this product is to supplement growth hormone levels in children who cannot make sufficient amounts of this compound. This allows these children to grow and develop normally.

However, in order to recoup their initial investment as quickly as possible, Genentech took an aggressive approach for expanding the market for this product (14). They funded front organizations, "charities," that gained access to elementary schools. These organizations screened children to identify those who were short of stature and referred them to local physicians for medical advice. To encourage sales, Genentech also rewarded physicians who frequently prescribed their growth hormone by funneling money to them, disguised as research grants or consulting fees. Some of these doctors even advertised their growth stimulating services on TV.

These tactics were very successful. In 1993 Genentech sold $225 million worth of human growth hormone to more than 14,000 clients. However, in August 1994, Genentech was indicted by federal prosecutors for these practices because large numbers of children—more than half of those treated—were treated with this hormone for cosmetic, not medical, reasons. The children were short not because they were deficient in growth hormone, but because they were naturally smaller.

These tactics might be acceptable for expanding the market for automobiles. However, they are not appropriate for marketing a hormone that profoundly influences the development of children, and misuse of which could seriously affect their health and well-being.

No research has been done to test whether the cosmetic use of growth hormone is safe. In patients who are clinically deficient in growth hormone, Genentech's product can cause cranial hypertension—high blood pressure localized to the brain, which can increase the risk of strokes. This side effect might be even more likely to occur when extra growth hormone is administered to children who already have normal levels. A number of other side effects could be expected in these children, as well. For instance, accelerating the growth process might adversely affect bone development, giving rise to problems such as defective or weak joints that could plague the individual throughout life.

The misuse of human growth hormone is not an isolated incident. Such tactics are common in the pharmaceutical industry. Commercial pressures often lead to diversion of medical technologies to non-medical uses which are potentially counterproductive to good health.

Applying this principle to human gene therapy, we can expect the same forces that expanded growth hormone sales from medical to cosmetic uses, to quickly expand the use of gene manipulations from therapeutic applications to use in genetic "enhancement." The dangers of such uses have been discussed at length earlier and need not be reiterated here. The point to be emphasized is that the imperatives of the business environment virtually guarantee that, if we allow genetic manipulations in humans for any reason, these abuses will become realities.

The balance of benefit to harm

It is noteworthy and unsettling that, within the context of current medical practices, the negative impacts described in the two examples above are generally considered to be within the range of acceptability. Drug companies routinely develop and market pharmaceuticals that have debilitating and even lethal side effects for some people. They know that doctors will prescribe those drugs because they have no better alternative; those drugs offer the best benefit-to-harm ratios of the products on the market.

This problem causes concern, but because the biomedical research community sees no alternative, it is simply accepted as being unavoidable. For instance, in evaluating a drug for migraine headaches, which has lethal side effects for a segment of the population, Dr. Paul Leber, Director of the FDA's Division of Neuropharmacological Drug Products, stated: "If there are to be potent drugs like sumatriptan . . . society must be willing to tolerate the injury that they will cause to some proportion of those who use them (15)."

In this environment, certain genetically engineered pharmaceuticals and diagnostics are viewed as advances in medical science. This style of reasoning has also been used to justify some applications of somatic cell gene therapy. However, these applications still carry real and substantial risks. In the final analysis, there is no good reason for accepting these risks. Instead of assuming that there are no better options, we should actively explore other approaches that show promise of more safely, simply, naturally, and cost-effectively addressing the needs that these genetic technologies attempt to fill.

Summary

We have seen that manipulating the human genome, especially the genome of germ-line cells, is hazardous and can be misused. We have also presented examples from the pharmaceutical industry suggesting that economic pressures will inevitably lead to abuses, if manipulation of the human germ line is allowed for medical purposes. We have also established that, in the final analysis, germ-line genetic alterations are not needed medically. In the absence of medical justification, the non-medical uses must stand on their own merit. The dangers of germ-line genetic manipulations can no longer be seen as an evil that must be risked out of medical necessity. Consequently, those who still wish to promote this technology must argue the validity of the remaining applications. Since these are cosmetic, social, and political in nature, it is highly unlikely that a well-informed public will accept such arguments.

CHAPTER 3

MAHARISHI'S VEDIC APPROACH TO HEALTH: FREEDOM FROM DISEASE THROUGH PREVENTION

Scientists are pragmatic. They are willing to adopt the best methods available for solving the problem at hand. In recent years, biomedical researchers have adopted the molecular biological approach, which includes genetic engineering, because they perceived this as most promising.

Although the resulting research has uncovered intriguing new knowledge concerning the molecular functioning of living systems, it has not widely succeeded in creating practical solutions to important diseases. For instance, molecular biologists have now discovered over 70 oncogenes (genes which, when mutated, can cause cancer); yet, the billions of dollars and thousands of researcher-years invested in this research to date have not resulted in a single practical measure for cancer treatment or prevention, nor has this approach led to breakthroughs for any of the other major health problems of today.

Despite this dismal track record, proponents claim that molecular medicine and genetic engineering are the hope of the future for solving the big health problems—heart disease, cancer, anxiety, depression, aging, drug abuse, and violent behavior. As we will see in this chapter, the logical approach to these problems is prevention, not high-tech therapeutics. We will also see that Maharishi's Vedic Approach to Health offers a range of holistic, natural-law-based methods that transforms the promise of prevention into a practical reality.

The Case For Prevention

The causes of almost all diseases are environmental and behavioral, not genetic

If we look for the root causes of the big health problems listed above, we find that environmental and behavioral contributions are by far the most important. These diseases are primarily due, not to defective genes, but to environmental influences or specific actions carried out by the individual. These influences and actions trigger the disease or contribute substantially to its development. When this is recognized, it becomes obvious that the aspiration of genetic engineers to solve these problems is naively unrealistic. How can we expect genetic approaches to solve health problems that are not caused by genetic defects?

Most lung cancer, for instance, is caused by a specific activity—smoking. Genetics may contribute to an individual's tendency to contract lung cancer, but that contribution is small, at most 10–15%. This means that genetic manipulations could, at best, reduce lung cancer by 10–15%. Furthermore, that small genetic contribution is due to the combined effects of a number of different genes. This makes genetic intervention even more impractical because it would be necessary to alter, not just one gene, but several.

The solution: simple, natural prevention, not high-tech genetic engineering

According to the U.S. Surgeon General, the majority of disease is self-induced and as much as 80% of medical problems can be prevented through behavioral or lifestyle change. In 1990, the U.S. Department of Health and Human Services published an exhaustive analysis of research on prevention, concluding that "better control of fewer than 10 health risk factors—for example, poor diet, infrequent exercise, use of tobacco and drugs, and abuse of alcohol—could prevent between 40–70% percent of all premature deaths, as well as a third of all cases of acute disability and two-thirds of all cases of chronic disability (16)." A recent study (17) confirmed this conclusion, showing that approximately 50% of all the deaths in the U.S. are premature and preventable. These premature deaths are due to the same behavioral and environmental factors mentioned above.

In another study, researchers empirically evaluated how treatment, prevention, information, and research contribute to health, as measured by life expectancy (18). These researchers concluded:

> Readily available empirical data suggest that until recent decades in the United States, and even today in nearly all underdeveloped nations, health improvement as measured by increased life expectancy has been almost entirely the result of improvements in prevention.

These research studies point to tremendous potential for improving health through disease prevention. Experience bears this out: educating people to avoid risk factors has already significantly reduced cardiovascular disease in the last 20 years and has been shown to be of some benefit in randomized clinical trials (19). Epidemiological evidence also bears this out: we could prevent 25% of all cancer deaths just by inspiring people to stop smoking (20).

A comprehensive prevention strategy is needed

Has the evidence cited above been translated into research on or implementation of effective strategies for prevention? The answer is "NO."

There are two reasons for this. First, there are tremendous economic pressures promoting the high-tech, therapeutic approach to medicine. The medical industry is one of the largest; in the U.S., it accounts for more than 14% of the Gross National Product. This approach is the bread and butter of physicians, the pharmaceutical industry, and all other aspects of the health care structure as it exists today. Medicine will always need therapeutic modalities, but the current focus on the treatment of disease—to the almost complete exclusion of preventive approaches—is not wise or balanced.

The second reason that prevention has not been implemented fully is because, even though it is easy for analysts to conclude from data on disease trends that most illness could be prevented, it is another thing to develop successful preventive approaches.

Public health measures, vaccination

In the latter part of the nineteenth and early part of the twentieth century, public health measures—such as improved sewage disposal and sanitation and improvements in the standard of living for the general population of Europe and North America—led to significant increases in longevity. Such improvements are still in progress in developing countries and are reflected in increasing life-span in those areas of the world.

Antibiotics and vaccination have also contributed to improving general health, although to a much lesser extent than is claimed by those who hold these up as the first great breakthroughs of modern medicine. It has been the simple hygiene measures that have actually made the big dent in infectious diseases.

Early detection

Another "preventive" approach popular today is early detection. For instance, sensitive mammographic methods are used to detect breast cancer in its early stages. Surgery is then used to remove malignant tissue before the tumor metastasizes. This approach is often effective, but prevention of cancer in the first place would obviously be preferable.

All of the approaches described above are useful preventive measures that continue to contribute to health. However, by themselves, they are not capable of solving the health problems of today. Moreover, current health problems such as cardiovascular disease, cancer, and drug abuse are much more challenging than the infectious diseases of the past.

Education and behavior modification

The preventive measures taken to address today's problems generally fall into two categories: educational measures and attempts to modify behaviors that are viewed as detrimental to health. At best, these approaches have been only partially successful. For instance, extensive efforts over the last 15 years have succeeded in educating the public regarding risk factors related to heart disease. People have learned that by improving diet and avoiding harmful behaviors such as smoking and alcohol consumption, they can improve their chances of avoiding heart disease. This has resulted in a 15% reduction of cardiovascular disease over this period of time. This is a great accomplishment that has saved millions of lives. However, this disease is still the biggest killer in industrialized nations, and the risk factors that contribute to this disease are still highly prevalent. Likewise, extensive efforts to educate the public about the dangers of smoking have led to small (10–15%) reductions in tobacco use and to small reductions in lung cancer. However, smoking-induced lung cancer is still the single largest category of cancer, responsible for 25% of all cancer deaths.

Lifestyle interventions, such as Johnson and Johnson's *Live for Life Program*, are another example of current prevention efforts. These attempt to comprehensively address all known important risk factors, and thereby develop and maintain healthier lifestyles. Programs of this type consist of specific interventions for each risk factor, such as smoking cessation, weight control, nutrition education, fitness, and blood pressure reduction programs. Researchers found that, over a five-year period, the medical expenses of those taking part in the *Live for Life Program* increased more slowly than those of control groups, but medical expenses still continued to rise (21). Thus, this and similar programs are of some benefit, but appear to improve health only to a small extent.

Current prevention strategies too superficial

As a result of the very limited success of past prevention efforts, prevention is looked on as a nice idea but not a very effective way to address serious health problems. The medical profession and the public fall back on biotechnology, and other high-tech biomedical approaches for magic bullets, that will rescue them from the results of health-eroding behavior. However, people's confidence in high-tech medicine is misplaced. It looks and sounds impressive, but it does not deliver the quick, easy, and effective cures promised by the magic bullet fantasy, and often creates as much pain and suffering as it relieves—or more. The false sense of security engendered by high-tech medicine is of little comfort when serious health problems arise.

The basic problem with the preventive approaches now in use is that they work on a level that is too superficial. Educational programs such as *Live for Life Program* are not powerful enough to change the behavior patterns and environmental conditions that cause disease. If external pressure is the primary influence in modifying behavior, then the deeper problems that initially gave rise to that behavior will manifest in some other way, causing other problems. The pressures that cause people to smoke often induce other harmful behaviors—overeating for example—if individuals are deprived, or deprive themselves, of tobacco. Alternatively, individuals may find themselves incapable of altering a certain behavior that they know is destructive to their health. We all know people who are perpetually trying to stop smoking or perpetually trying to modify their eating habits to lose weight. Instead of improving health, such efforts simply degrade the self-esteem of the individual.

Deeper Knowledge Needed

What we really need is deeper knowledge to empower prevention. The idea behind comprehensive lifestyle interventions is a good one, but *Live for Life* and similar programs do not go far enough. They are comprehensive in the sense that they take into account all of the risk factors for important diseases that research has identified to date, but such a list is not truly comprehensive, and never can be. It is impossible to experimentally identify all of the risk factors for all diseases, and even if they all could be identified, it would be impossible to design a lifestyle intervention that could incorporate measures for correcting all of those risk factors. Furthermore, for interventions to be successful, they need to work from a level much more fundamental than education or behavioral manipulation.

The strategy of addressing risk factors one-by-one is a natural outgrowth of the objective, reductionistic scientific approach, upon which modern medicine relies. A truly comprehensive approach to prevention must be based on a much more holistic approach to gaining knowledge. Ayur-Veda, the traditional medical system of India, is based on such knowledge. In contrast to the modern, treatment-oriented medical system, this system is prevention oriented and holistic. This system does not try to manipulate individual risk factors, one-by-one. Rather, it takes a comprehensive and integrated approach whose objective is to maintain or, when necessary, to re-establish balance in all aspects of the life of the individual, thereby avoiding disease at its inception—in one word: prevention.

If we intend to devise a comprehensive prevention strategy for contemporary society, it makes good sense to use the knowledge of prevention that has been preserved in Ayur-Veda. Why waste time rediscovering the wheel, when this time-tested body of knowledge is available?

Maharishi's Vedic Approach to Health
A safe, effective approach for prevention and treatment of disease

Until about 15 years ago, Ayur-Veda was inaccessible to Western medicine and was in disrepair, due to many centuries of cultural, economic, and political upheaval. In the early 1980s, Maharishi Mahesh Yogi—the leading authority on the Vedic system of knowledge—began working to restore Ayur-Veda, with support from Ayur-Vedic physicians, Vedic scholars, and Western physicians and scientists. The result of this work is Maharishi's Vedic Approach to Health, or Maharishi Ayur-Veda. In this system, we find Ayur-Veda restored to its original integrity. In Sanskrit, *Ayu* means life, and *Veda* means knowledge. Maharishi Ayur-Veda, Maharishi's Vedic Approach to Health, is a knowledge-based system of health care that provides comprehensive, systematic, and effective prevention and therapeutics.

Scientific evidence for the effectiveness of Maharishi's Vedic Approach to Health

Considerable research has already shown that the many modalities of Maharishi's Vedic Approach to Health are effective in preventing and treating disease. Over 500 research studies on this approach have been published by leading scientists from over 200 different research institutions around the world (22). This body of research establishes the effectiveness of this natural but powerful program.

Among this body of data is research that clearly documents the fact that Maharishi's Vedic Approach to Health successfully addresses the major diseases that confront society today, including the big killers, cardiovascular disease and cancer.

For instance, extensive research over the past 15 years, including research funded by the U.S. National Institutes of Health, has established that this approach is highly effective in treating hypertension and preventing cardiovascular disease. One recent study, funded by the U.S. National Institutes of Health, demonstrated that

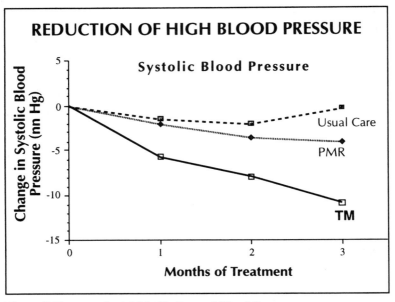

Figure 7. Transcendental Meditation and Blood Pressure
Elderly inner-city African-Americans with moderately elevated blood pressure were randomly assigned to groups practicing Transcendental Meditation, progressive muscle relaxation (PMR), which is a common relaxation technique, or usual care. Over a three-month interval, systolic and diastolic blood pressure dropped by 10.6 and 5.9 mm Hg in the Transcendental Meditation group, and 4.0 and 2.1 mm Hg in the PMR group, with virtually no change in the usual care group. Also, the Transcendental Meditation group showed larger improvements than other groups on multiple quality of life indicators such as decreased health complaints. A second random assignment study with the elderly, conducted at Harvard University, found similar systolic blood pressure changes produced by Transcendental Meditation over three months (11 mm Hg for systolic blood pressure). Chapter 14 in *Personality, Elevated Blood Pressure, and Essential Hypertension* pp. 291–316 (Washington, DC: Hemisphere Publishing, 1992); *Journal of Personality and Social Psychology*, 57 (1989): 950–964.

this approach was highly effective in treating hypertension in an urban population of elderly African-Americans (23). The results of this study are presented in Figure 7. Researchers from Maharishi International University are currently directing a large study, supported by a $2.5 million grant from the National Heart, Lung and Blood Institute, the aim of which aim is to expand and extend their earlier work.

Another very interesting study (24) showed that one of the primary modalities of the Vedic approach, Maharishi's Transcendental Meditation technique, is effective in preventing a wide range of diseases. Using data obtained from the Blue Cross/Blue Shield insurance company, researchers found, as shown in Figure 8, that individuals trained to practice this technique needed far less medical attention, as measured by hospital admission rates and doctors' visits. Looking at hospital admission rates for 16 different categories of disease, the researchers found that those who practiced the Transcendental Meditation technique were healthier across the board. As shown in Figure 9, hospital admission rates were significantly reduced in all sixteen categories.

These results have been confirmed by a more recent study (25), that used data obtained from the government health insurance agency of the Province of Quebec, Canada. Because of the detailed medical information collected by this state-run program, researchers were able to follow health care utilization patterns over a period of several years. This study compared the health care utilization patterns during the three years before an individual started Transcendental Meditation to the utilization patterns during the three to seven years immediately following instruction in this technique. They found that health care utilization was constant during the three years before starting and then dropped by 7% during each

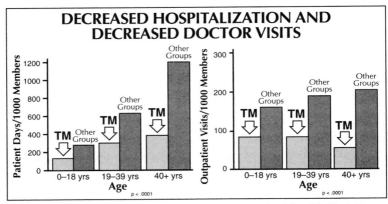

Figure 8. Transcendental Meditation Reduces the Need for Medical Care
A study of health insurance statistics on over 2000 people practicing Maharishi's Transcendental Meditation and TM-Sidhi Program over a 5-year period found that participants required much less medical care than other groups matched for age, gender, profession and insurance terms. Those who practiced Transcendental Meditation required less than half as many doctor visits and days of hospitalization, and the difference between Transcendental Meditation meditators and control groups increased in older age-brackets. *Psychosomatic Medicine*, 49 (1987): 493-507.

---

year after beginning the practice. Thus, practice of Transcendental Meditation reduced health care utilization by 21% by the end of the third year, by 35% by the end of the fifth year, and by 49% by the end of the seventh year. This indicates that this technique greatly improves health.

Interestingly, these improvements were observed, even for individuals who were initially categorized as "high health care utilizers." These are the 5% of the population who use 70% of our health care resources—the very ill. The effectiveness of this program with this group indicates that Transcendental Meditation is very effective therapeutically, as well as preventively.

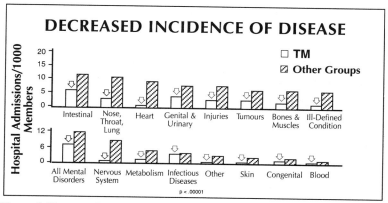

Figure 9. Transcendental Meditation Reduces Disease in All Major Categories
The study described in Figure 8 also examined specific categories of disease, and found that disease incidence decreased in all sixteen categories examined, including large decreases in major diseases such as heart disease and cancer. *Psychosomatic Medicine*, 49 (1987): 493-507.

The categories of disease included in the Canadian study were similar to those for the earlier Blue Cross/Blue Shield study (Figures 8 and 9). Substantial improvements were observed in all major categories, including heart disease, cancer, infectious diseases, mental disorders, and diseases of the nervous system.

If scientists at the National Institutes of Health or some other well known biomedical research institution were to discover a new drug that could reduce occurrence of even one of these diseases to the extent that Maharishi's Vedic Approach to Health does, it would be proclaimed on the front page of every important newspaper in the world as a major medical breakthrough. Here we see that this one technology provides breakthrough-quality benefits in all of these areas.

Expanding research on Maharishi's Vedic Approach to Health

Today, the pace of research on Maharishi's Vedic Approach to Health continues to accelerate. Research programs in progress at this time include the following:

- clinical studies of Vedic approaches to treating hypertension, coronary artery disease, AIDS, adult-onset diabetes mellitus, rheumatoid arthritis, bronchial asthma, migraine headache, and irritable bowel syndrome;
- effects of the dietary regimens of Maharishi's Vedic Approach to Health on blood lipids and coronary heart disease;
- use of Maharishi Ayur-Veda herbal preparations in treating atherosclerosis;
- effects of physiological purification therapies of Maharishi's Vedic Approach to Health on blood lipids and immune parameters;
- anti-cancer effects of Maharishi Ayur-Veda herbal preparations;
- use of Maharishi's Vedic Approach to Health to reduce toxicity in cancer chemotherapy;
- effects of the herbal supplement Maharishi Amrit Kalash on aging, DNA repair, and scavenging of free radicals.

The central principle

Why is Maharishi's Vedic Approach to Health so effective? This can best be understood by comparing the underlying principle of this system with those of modern medicine.

Modern medicine

In order to understand the physiology, modern biomedical science has artificially segmented the whole into parts and studied those parts in isolation. Interesting knowledge has come out of this exercise. However, in the process, scientists and physicians have forgotten that the reality is the whole, and that in real-world medicine the physiology must be dealt with as an integrated, unified whole.

The approach of modern medicine is fundamentally materialistic and mechanistic. It views the body as a machine, and views the patient in much the same way that an automobile mechanic views a damaged auto: find that part of the machine that is broken and fix it. Genetic engineering takes this logic to its ultimate extreme. It takes the perspective that every structure and function of the body is specified by a gene or set of genes; everything has a genetic blueprint. Therefore, the ultimate remedy for any disease will be to "fix" the gene for whatever part of the physiology is "broken." However, this is a flawed vision of human life and human health. Health is not a function of the properties of isolated components of the physiology, but rather of the balanced, integrated functioning of the whole. Even if we were to genetically re-engineer ten genes, or even one hundred, we could not succeed in reprogramming the body to be healthy.

By attempting to deal with health piecemeal, medicine has been reduced to an unending exercise in correcting symptoms. An initial therapeutic intervention corrects one problem, or part of it, but leads to other side effects, which in turn must be corrected, either in the long or short term, by another intervention. This second intervention brings with it its own side effects, which must also be corrected, and so on, *ad infinitum.* This approach is clearly not ideal. It is not even adequate, and physicians know it. However without a broader vision, they are trapped in this endless series of mistakes.

The basic problem with this approach is that it is too superficial. It attempts to exploit knowledge of isolated laws of nature in order to maintain health; but health is an integrative, holistic property. This superficial approach is problematic even with routine pharmacological and surgical remedies, but is much more of a problem with genetic interventions because these influence such a fundamental level of the physiology.

To be successful in maintaining and restoring health, medicine must deal with the human being as a whole. To do this, medical practice must work at a level that is common to and integrates all aspects of the physiology. Currently, medicine lacks a viable strategy for accomplishing this. Maharishi's Vedic Approach to Health offers such methods.

Maharishi's Vedic Approach to Health

In the Introduction we briefly discussed the idea that the structure and orderliness of nature are expressions of an underlying field of intelligence. We established that technologies that work on the level of this field of intelligence are capable of addressing problems in any area of life thoroughly, comprehensively, and without creating harmful side effects.

Maharishi's Vedic Approach to Health, like all of the other technologies of Maharishi's Vedic Engineering, are based on knowledge of this inner intelligence, the unified field of natural law. According to this system of medicine, the laws of nature that direct the functioning of the physiology are localized, individualized expressions of the fundamental, unifying level of nature's intelligence.

The central strategy of this approach is to work from this holistic level of intelligence in order to integrate the functioning of all of the localized expressions of natural law. This allows all aspects of the physiology to function in an integrated and balanced

Figure 10—Physiology and the Unified Field (pages 76 and 77)

This chart illustrates how Maharishi's Vedic Science provides a new, integrated approach in which the whole range of physiology can be appreciated from its source in the unified field of all the laws of nature. The left-hand side shows how this field gives rise to the fundamental force and matter fields, the basis of all physical-chemical transformations, expressed in the physiology as the DNA—the material expression of the unified field that creates and governs the flow of biological intelligence on all levels of physiological organization.

The sequential expression of knowledge and organizing power from the unified field is displayed in this chart in terms of six hierarchical levels: Level 1—the unified field; Level 2—the fundamental force and matter fields; Level 3—the DNA and RNA molecules; Level 4—the expressed levels of physiological organization (proteins, biochemical pathways, cell components, cells, tissues, organs, and organ systems); Level 5—individual physiology and consciousness; and Level 6—the physiology and collective consciousness of society.

The right-hand side illustrates that, through Maharishi's Transcendental Meditation and TM-Sidhi program, the individual can experience the unified field of all the laws of nature as the simplest state of awareness, the self-referral state of transcendental consciousness. Direct experience of the unified field gives the individual access to the most fundamental level of natural law, referred to as *Veda* in Maharishi's Vedic Science. Veda is the field of pure intelligence and is the source of all the impulses of natural law that structure creation.

These impulses of natural law constitute the dynamic principles that govern the universe. The familiarity with these principles gained through Maharishi's Transcendental Meditation and TM-Sidhi program is not intellectual understanding in the ordinary sense. Instead, individual awareness identifies with the mechanics through which natural law expresses itself in its own self-referral state. Established in this self-referral state of wholeness, every impulse of thought and behavior is an impulse of natural law. In this state, the behavior of the individual and society is brought into accord with the total potential of natural law and is therefore completely evolutionary and life supporting.

PHYSIOLOGY A

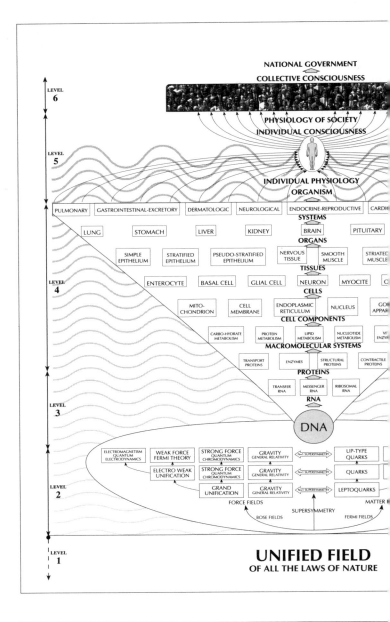

MAHARISHI'S VEDIC APPROACH TO HEALTH

E UNIFIED FIELD

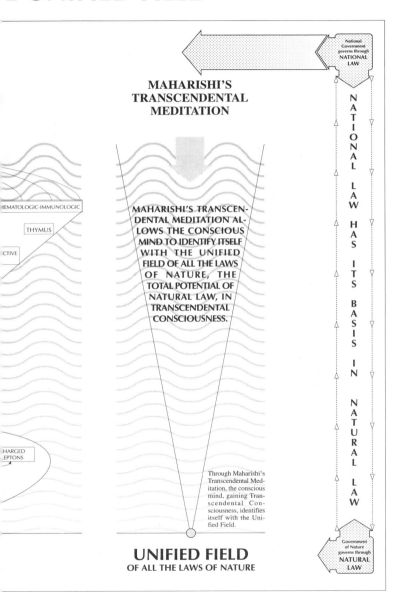

Through Maharishi's Transcendental Meditation, the conscious mind, gaining Transcendental Consciousness, identifies itself with the Unified Field.

manner, restoring balance in the life of the patient. That balance is the essence of health.

Nature's intelligence: the unified field of natural law

What is this holistic level of nature's intelligence? Both modern science and Maharishi's Vedic Science offer relevant insights. Over the centuries, modern science has gradually elucidated the hierarchical organization of nature, depicted in the left-hand portion of Figure 10. There we see a continuum, extending from the force and matter fields of quantum physics, to the atoms and molecules of chemistry, to the biochemical pathways and organelles of the cell, to the tissues and organs of the complete physiology, and finally to the social structures and the biosphere of which each individual organism is a part.

Quantum physicists have recently contributed something new to this scheme. They have discovered the unified field of all the laws of nature—the ultimate source of order in the universe. According to these new theories of quantum physics, the unified field is the single source from which all of nature emerges.

The right-hand side of Figure 10 illustrates how direct experiential knowledge of this field is gained through the primary technologies of Maharishi's Vedic Science, the Transcendental Meditation technique and the more advanced TM-Sidhi program. Transcendental Meditation enables the individual to experience progressively less active, less excited levels of thought, and finally to transcend thought completely. In this transcendental state, one is awake inside but there are no thoughts, no thinking activity. It is a state of restful alertness, in which the mind is awake—even more so than usual—but free of thoughts, silent. Awareness is not local-

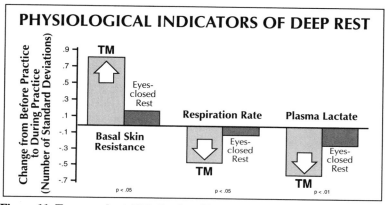

Figure 11. Transcendental Meditation Gives Deep Rest
When practicing Maharishi's Transcendental Meditation technique, one gains rest that is much deeper than that experienced when sitting quietly with eyes closed. This chart shows that basal skin resistance increases, while blood lactate falls during practice of Transcendental Meditation, indicating a deep state of psycho-physiological relaxation. In parallel, respiration rate drops, indicating reduced metabolism and physiological rest. These results were obtained in a meta-analysis of more than thirty published research reports examining the physiological effects of Transcendental Meditation. Meta-analysis is a procedure that enables the researcher to draw definitive conclusions from large bodies of data obtained from several research studies. *American Psychologist* 42 (1987): 879-881.

---

ized to isolated thoughts, but is expanded without limits. This is unbounded awareness, a state of pure consciousness.

Extensive research has shown that this mental state is accompanied by distinct and reproducible physiological changes. As illustrated in Figure 11, the body gains a state of deep rest during practice of the Transcendental Meditation technique; yet the brain takes on a more orderly, coherent, and integrated style of functioning (Figure 12). Brain physiologists point out that this state of restful alertness is unique to the practice of the Transcendental Meditation technique.

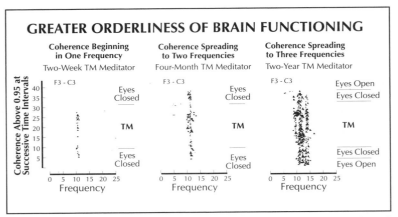

Figure 12. Transcendental Meditation Generates Orderly Functioning of the Brain
EEG coherence increases during practice of the Transcendental Meditation technique, indicating greater orderliness of brain functioning. Other studies have shown that increased levels of EEG coherence measured during the practice are significantly correlated with improvements in physiological and mental functioning, including increased creativity. This chart illustrates the progressive enhancement of brain functioning that occurs as one practices this technique over a period of weeks, months, and years. It shows that, for a two-week meditator, increased brain wave coherence is induced during the practice, and then continues for a few minutes afterwards. After four months of meditation, coherence is present even immediately before practicing the technique, spreads to a wider range of frequencies during the practice, and continues after meditation. For people who have practiced Transcendental Meditation regularly over longer periods of time (in this chart, for two years), coherence becomes apparent as soon as eyes are closed, increases in power, spreads to a wider range of frequencies during the practice, and continues after meditation even when eyes are open. *Psychosomatic Medicine*, 46 (1984): 267-276; *San Diego Biomedical Symposium*, 15 (1976).

Through the technologies of Maharishi's Vedic Science, experiential familiarity with pure consciousness is gained, revealing it to be an unbounded field of energy, creativity, and intelligence, from which emerge all of the isolated laws of nature that govern all localized phenomena of life (see Figure 10, pages 75–77).

Thus, as Maharishi Mahesh Yogi says in his book, *Maharishi's Absolute Theory of Defense* (26), "Both understandings, modern and ancient, locate the unified source of Nature's perfect order in a single, self-interacting field of intelligence at the foundation of all the Laws of Nature. This field sequentially creates from within itself all the diverse Laws of Nature governing life at every level of the manifest universe."

Using their own terminologies, modern physics and Maharishi's Vedic Science both describe the dynamics through which all of the localized laws of nature emerge from the unified field of natural law. In quantum field theory this description is expressed in the form of a highly complex mathematical formula, the D=10 Lagrangian of the unified field, shown in the right-hand side of Figure 13. An analogous, and equally precise and systematic statement in the notation of the Vedic system is presented in the left-hand side of Figure 13. Thus both systems of knowledge uphold each other in providing a basic understanding of the unified field of natural law.

Maharishi's Vedic Science goes a step beyond modern science, however. It provides technologies—Maharishi's Vedic

Figure 13. Vedic and Modern Descriptions of the Unified Field of All the Laws of Nature *(overleaf).*
Both Maharishi's Vedic Science and modern physics have identified a single, universal source of all orderliness in nature—the unified field of all the laws of nature. In addition, Maharishi's Vedic Science provides a practical, scientifically validated procedure to apply this most fundamental and powerful level of natural law for the benefit of humanity. Both systems describe this field in their own terminologies and describe how this field sequentially creates, from within itself, all the diverse laws of nature governing life at every level of the manifest universe. This figure presents these two complementary descriptions. A more detailed discussion of this figure is presented in Appendix II under the title Constitution of the Universe.

Constitutio...

Vedic Science

Ahaṁkār	Buddhi	Manas	Ākāsh	Vāyu	Agni	Jal	Prithivī
ऋक्	नि	मी	ळे	पु	रो	हि	तं
AK	NI	MI	LE	PU	RO	HI	TAM
त्र	ग्निः	पू	वँ	भिः	ऋ	षि	भि
त्र	ग्नि	नो	र्	यि	मं	श्न	वृं
त्र	ग्ने	यं	य्	ज्ञ	मं	ध्व	रं
त्र	ग्निर्	हो	ता	क्	वि	क्रं	तुः
य	द्	ह्ं	दा	शु	षे	तु	वं
उ	प	त्वा	ग्ने	दि	वे	दि	वे
रा	ज	न्त	म	ध्व	रा	शां	गो
स	नः	पि	ते	वं	सू	न	वे

Ahaṁkār	Buddhi	Manas	Ākāsh	Vāyu	Agni	Jal	Prithivī
य	ज्ञ	स्यं	दे	व	मृ	त्वि	जम्
YA	GYA	SYA	DE	VA	MRI	TVI	JAM
री	ड	यो	नू	तं	नै	रु	त
त्यो	ष	मे	व	दि	वे	दि	वे
वि	श्व	तः	प	रि	भू	र	सि
सं	त्यश्	चि	त्र	श्रं	व	स्त	मः
त्र	मैं	भ	द्रं	कं	रि	ष्य	सि
दो	षा	व	स्तर्	धि	या	व्	यम्
पा	मृ	त	स्यु	दीं	दि	वि	म्
त्र	ग्रे	सू	पा	यु	नो	भ	व

Ahaṁkār	Buddhi	Manas	Ākāsh	Vāyu	Agni	Jal	Prithivī
हो	ता	रं	र	त्न	धा	तं	मम्
HO	TA	RAM	RA	TNA	DHA	TA	MAM
स	दे	वाँ	ए	ह	वं	त्त	ति
य	श	सं	वी	र	वं	त्त	मम्
स	इ	हे	वे	षुं	ग	च्छ	ति
दे	वो	दे	वे	भिं	रा	गं	मत्
त	वेत्	तत्	सं	त्य	मं	ह्नि	रः
न	मो	ध	रं	नं	ए	मं	सि
व	धं	मा	नं	सु	वे	द	में
स	च	सु	आ	नः	स्व	स्त	ये

e Universe

Modern Science

$$L_F^{(10)} = \tfrac{1}{\pi}(\psi_L^1\partial_+\psi_L^1 + \psi_L^2\partial_+\psi_L^2 + \psi_L^3\partial_+\psi_L^3 + \psi_L^4\partial_+\psi_L^4 + \psi_L^5\partial_+\psi_L^5 + \psi_L^6\partial_+\psi_L^6 + \psi_L^7\partial_+\psi_L^7 + \psi_L^8\partial_+\psi_L^8)$$

$$\begin{aligned}L_F^{(4)} =\ & \tfrac{1}{\pi}(\overline{\Psi}_L^1\partial_+\overline{\Psi}_L^1 + \overline{\Psi}_L^2\partial_+\overline{\Psi}_L^2 + \chi_L^1\partial_+\chi_L^1 + \chi_L^2\partial_+\chi_L^2 + \chi_L^3\partial_+\chi_L^3 + \chi_L^4\partial_+\chi_L^4 + \chi_L^5\partial_+\chi_L^5 + \chi_L^6\partial_+\chi_L^6 \\ &+ y_L^1\partial_+y_L^1 + y_L^2\partial_+y_L^2 + y_L^3\partial_+y_L^3 + y_L^4\partial_+y_L^4 + y_L^5\partial_+y_L^5 + y_L^6\partial_+y_L^6 + \omega_L^1\partial_+\omega_L^1 + \omega_L^2\partial_+\omega_L^2 \\ &+ \omega_L^3\partial_+\omega_L^3 + \omega_L^4\partial_+\omega_L^4 + \omega_L^5\partial_+\omega_L^5 + \omega_L^6\partial_+\omega_L^6 + \overline{y}_R^1\partial_-\overline{y}_R^1 + \overline{y}_R^2\partial_-\overline{y}_R^2 + \overline{y}_R^3\partial_-\overline{y}_R^3 + \overline{y}_R^4\partial_-\overline{y}_R^4 \\ &+ \overline{y}_R^5\partial_-\overline{y}_R^5 + \overline{y}_R^6\partial_-\overline{y}_R^6 + \overline{\omega}_R^1\partial_-\overline{\omega}_R^1 + \overline{\omega}_R^2\partial_-\overline{\omega}_R^2 + \overline{\omega}_R^3\partial_-\overline{\omega}_R^3 + \overline{\omega}_R^4\partial_-\overline{\omega}_R^4 + \overline{\omega}_R^5\partial_-\overline{\omega}_R^5 + \overline{\omega}_R^6\partial_-\overline{\omega}_R^6 \\ &+ \psi_R^1\partial_-\psi_R^1 + \psi_R^2\partial_-\psi_R^2 + \psi_R^3\partial_-\psi_R^3 + \psi_R^4\partial_-\psi_R^4 + \psi_R^5\partial_-\psi_R^5 + \eta_R^1\partial_-\eta_R^1 + \eta_R^2\partial_-\eta_R^2 + \eta_R^3\partial_-\eta_R^3 \\ &+ \overline{\psi}_R^1\partial_-\overline{\psi}_R^1 + \overline{\psi}_R^2\partial_-\overline{\psi}_R^2 + \overline{\psi}_R^3\partial_-\overline{\psi}_R^3 + \overline{\psi}_R^4\partial_-\overline{\psi}_R^4 + \overline{\psi}_R^5\partial_-\overline{\psi}_R^5 + \overline{\eta}_R^1\partial_-\overline{\eta}_R^1 + \overline{\eta}_R^2\partial_-\overline{\eta}_R^2 + \overline{\eta}_R^3\partial_-\overline{\eta}_R^3 \\ &+ \phi_R^1\partial_-\phi_R^1 + \phi_R^2\partial_-\phi_R^2 + \phi_R^3\partial_-\phi_R^3 + \phi_R^4\partial_-\phi_R^4 + \phi_R^5\partial_-\phi_R^5 + \phi_R^6\partial_-\phi_R^6 + \phi_R^7\partial_-\phi_R^7 + \phi_R^8\partial_-\phi_R^8 \\ &+ \overline{\phi}_R^1\partial_-\overline{\phi}_R^1 + \overline{\phi}_R^2\partial_-\overline{\phi}_R^2 + \overline{\phi}_R^3\partial_-\overline{\phi}_R^3 + \overline{\phi}_R^4\partial_-\overline{\phi}_R^4 + \overline{\phi}_R^5\partial_-\overline{\phi}_R^5 + \overline{\phi}_R^6\partial_-\overline{\phi}_R^6 + \overline{\phi}_R^7\partial_-\overline{\phi}_R^7 + \overline{\phi}_R^8\partial_-\overline{\phi}_R^8)\end{aligned}$$

$$L_F^{(10)} = \tfrac{1}{\pi}(\psi_L^1\partial_+\psi_L^1 + \psi_L^2\partial_+\psi_L^2 + \psi_L^3\partial_+\psi_L^3 + \psi_L^4\partial_+\psi_L^4 + \psi_L^5\partial_+\psi_L^5 + \psi_L^6\partial_+\psi_L^6 + \psi_L^7\partial_+\psi_L^7 + \psi_L^8\partial_+\psi_L^8)$$

$$\begin{aligned}L_F^{(4)} =\ & \tfrac{1}{\pi}(\overline{\Psi}_L^1\partial_+\overline{\Psi}_L^1 + \overline{\Psi}_L^2\partial_+\overline{\Psi}_L^2 + \chi_L^1\partial_+\chi_L^1 + \chi_L^2\partial_+\chi_L^2 + \chi_L^3\partial_+\chi_L^3 + \chi_L^4\partial_+\chi_L^4 + \chi_L^5\partial_+\chi_L^5 + \chi_L^6\partial_+\chi_L^6 \\ &+ y_L^1\partial_+y_L^1 + y_L^2\partial_+y_L^2 + y_L^3\partial_+y_L^3 + y_L^4\partial_+y_L^4 + y_L^5\partial_+y_L^5 + y_L^6\partial_+y_L^6 + \omega_L^1\partial_+\omega_L^1 + \omega_L^2\partial_+\omega_L^2 \\ &+ \omega_L^3\partial_+\omega_L^3 + \omega_L^4\partial_+\omega_L^4 + \omega_L^5\partial_+\omega_L^5 + \omega_L^6\partial_+\omega_L^6 + \overline{y}_R^1\partial_-\overline{y}_R^1 + \overline{y}_R^2\partial_-\overline{y}_R^2 + \overline{y}_R^3\partial_-\overline{y}_R^3 + \overline{y}_R^4\partial_-\overline{y}_R^4 \\ &+ \overline{y}_R^5\partial_-\overline{y}_R^5 + \overline{y}_R^6\partial_-\overline{y}_R^6 + \overline{\omega}_R^1\partial_-\overline{\omega}_R^1 + \overline{\omega}_R^2\partial_-\overline{\omega}_R^2 + \overline{\omega}_R^3\partial_-\overline{\omega}_R^3 + \overline{\omega}_R^4\partial_-\overline{\omega}_R^4 + \overline{\omega}_R^5\partial_-\overline{\omega}_R^5 + \overline{\omega}_R^6\partial_-\overline{\omega}_R^6 \\ &+ \psi_R^1\partial_-\psi_R^1 + \psi_R^2\partial_-\psi_R^2 + \psi_R^3\partial_-\psi_R^3 + \psi_R^4\partial_-\psi_R^4 + \psi_R^5\partial_-\psi_R^5 + \eta_R^1\partial_-\eta_R^1 + \eta_R^2\partial_-\eta_R^2 + \eta_R^3\partial_-\eta_R^3 \\ &+ \overline{\psi}_R^1\partial_-\overline{\psi}_R^1 + \overline{\psi}_R^2\partial_-\overline{\psi}_R^2 + \overline{\psi}_R^3\partial_-\overline{\psi}_R^3 + \overline{\psi}_R^4\partial_-\overline{\psi}_R^4 + \overline{\psi}_R^5\partial_-\overline{\psi}_R^5 + \overline{\eta}_R^1\partial_-\overline{\eta}_R^1 + \overline{\eta}_R^2\partial_-\overline{\eta}_R^2 + \overline{\eta}_R^3\partial_-\overline{\eta}_R^3 \\ &+ \phi_R^1\partial_-\phi_R^1 + \phi_R^2\partial_-\phi_R^2 + \phi_R^3\partial_-\phi_R^3 + \phi_R^4\partial_-\phi_R^4 + \phi_R^5\partial_-\phi_R^5 + \phi_R^6\partial_-\phi_R^6 + \phi_R^7\partial_-\phi_R^7 + \phi_R^8\partial_-\phi_R^8 \\ &+ \overline{\phi}_R^1\partial_-\overline{\phi}_R^1 + \overline{\phi}_R^2\partial_-\overline{\phi}_R^2 + \overline{\phi}_R^3\partial_-\overline{\phi}_R^3 + \overline{\phi}_R^4\partial_-\overline{\phi}_R^4 + \overline{\phi}_R^5\partial_-\overline{\phi}_R^5 + \overline{\phi}_R^6\partial_-\overline{\phi}_R^6 + \overline{\phi}_R^7\partial_-\overline{\phi}_R^7 + \overline{\phi}_R^8\partial_-\overline{\phi}_R^8)\end{aligned}$$

$$L_F^{(10)} = \tfrac{1}{\pi}(\psi_L^1\partial_+\psi_L^1 + \psi_L^2\partial_+\psi_L^2 + \psi_L^3\partial_+\psi_L^3 + \psi_L^4\partial_+\psi_L^4 + \psi_L^5\partial_+\psi_L^5 + \psi_L^6\partial_+\psi_L^6 + \psi_L^7\partial_+\psi_L^7 + \psi_L^8\partial_+\psi_L^8)$$

$$\begin{aligned}L_F^{(4)} =\ & \tfrac{1}{\pi}(\overline{\Psi}_L^1\partial_+\overline{\Psi}_L^1 + \overline{\Psi}_L^2\partial_+\overline{\Psi}_L^2 + \chi_L^1\partial_+\chi_L^1 + \chi_L^2\partial_+\chi_L^2 + \chi_L^3\partial_+\chi_L^3 + \chi_L^4\partial_+\chi_L^4 + \chi_L^5\partial_+\chi_L^5 + \chi_L^6\partial_+\chi_L^6 \\ &+ y_L^1\partial_+y_L^1 + y_L^2\partial_+y_L^2 + y_L^3\partial_+y_L^3 + y_L^4\partial_+y_L^4 + y_L^5\partial_+y_L^5 + y_L^6\partial_+y_L^6 + \omega_L^1\partial_+\omega_L^1 + \omega_L^2\partial_+\omega_L^2 \\ &+ \omega_L^3\partial_+\omega_L^3 + \omega_L^4\partial_+\omega_L^4 + \omega_L^5\partial_+\omega_L^5 + \omega_L^6\partial_+\omega_L^6 + \overline{y}_R^1\partial_-\overline{y}_R^1 + \overline{y}_R^2\partial_-\overline{y}_R^2 + \overline{y}_R^3\partial_-\overline{y}_R^3 + \overline{y}_R^4\partial_-\overline{y}_R^4 \\ &+ \overline{y}_R^5\partial_-\overline{y}_R^5 + \overline{y}_R^6\partial_-\overline{y}_R^6 + \overline{\omega}_R^1\partial_-\overline{\omega}_R^1 + \overline{\omega}_R^2\partial_-\overline{\omega}_R^2 + \overline{\omega}_R^3\partial_-\overline{\omega}_R^3 + \overline{\omega}_R^4\partial_-\overline{\omega}_R^4 + \overline{\omega}_R^5\partial_-\overline{\omega}_R^5 + \overline{\omega}_R^6\partial_-\overline{\omega}_R^6 \\ &+ \psi_R^1\partial_-\psi_R^1 + \psi_R^2\partial_-\psi_R^2 + \psi_R^3\partial_-\psi_R^3 + \psi_R^4\partial_-\psi_R^4 + \psi_R^5\partial_-\psi_R^5 + \eta_R^1\partial_-\eta_R^1 + \eta_R^2\partial_-\eta_R^2 + \eta_R^3\partial_-\eta_R^3 \\ &+ \overline{\psi}_R^1\partial_-\overline{\psi}_R^1 + \overline{\psi}_R^2\partial_-\overline{\psi}_R^2 + \overline{\psi}_R^3\partial_-\overline{\psi}_R^3 + \overline{\psi}_R^4\partial_-\overline{\psi}_R^4 + \overline{\psi}_R^5\partial_-\overline{\psi}_R^5 + \overline{\eta}_R^1\partial_-\overline{\eta}_R^1 + \overline{\eta}_R^2\partial_-\overline{\eta}_R^2 + \overline{\eta}_R^3\partial_-\overline{\eta}_R^3 \\ &+ \phi_R^1\partial_-\phi_R^1 + \phi_R^2\partial_-\phi_R^2 + \phi_R^3\partial_-\phi_R^3 + \phi_R^4\partial_-\phi_R^4 + \phi_R^5\partial_-\phi_R^5 + \phi_R^6\partial_-\phi_R^6 + \phi_R^7\partial_-\phi_R^7 + \phi_R^8\partial_-\phi_R^8 \\ &+ \overline{\phi}_R^1\partial_-\overline{\phi}_R^1 + \overline{\phi}_R^2\partial_-\overline{\phi}_R^2 + \overline{\phi}_R^3\partial_-\overline{\phi}_R^3 + \overline{\phi}_R^4\partial_-\overline{\phi}_R^4 + \overline{\phi}_R^5\partial_-\overline{\phi}_R^5 + \overline{\phi}_R^6\partial_-\overline{\phi}_R^6 + \overline{\phi}_R^7\partial_-\overline{\phi}_R^7 + \overline{\phi}_R^8\partial_-\overline{\phi}_R^8)\end{aligned}$$

Engineering—that enable the individual to directly experience this field. Along with experiential knowledge of this fundamental level of natural law comes the ability to make full use of it to bring about integrated, balanced, side effect-free improvements in life. The following sections discuss those technologies in the context of health.

Discovery of Veda in human physiology

The unified field of natural law, or the field of pure consciousness, is relevant not only to physics but also to the life sciences and health. It underlies and is the most fundamental structure of every level of existence, including the human physiology. Therefore, direct experience of this level of natural law results in holistic knowledge of the fundamental laws governing the human physiology.

Recently, important progress in research on Maharishi's Vedic Approach to Health has been made by Dr. Tony Nader, M.D., Ph.D., International President of Maharishi Ayur-Veda Universities, who is a Massachusetts Institute of Technology and Harvard-trained neuroscientist. His research establishes fundamental connections between the field of pure consciousness and the physiology. He presents his discoveries in detail in his book entitled *Human Physiology—Expression of Veda and the Vedic Literature* (27).

In essence, Dr. Nader has discovered that the structure and function of the human physiology, as described by modern physiologists, correspond precisely to the internal structure and dynamics of the field of pure consciousness, or Veda, as displayed in the technical literature of Maharishi's Vedic Science, the Vedic literature.

Modern scientific literature records knowledge generated through the objective methods of experimentation and empirical observation. Similarly, the Vedic literature displays knowledge generated using systematic, Vedic approaches (introduced in the previous section), which provide highly reliable, direct experiential knowledge of the field of consciousness, or Veda.

The 27 branches of the Vedic literature display the internal structure and dynamics of consciousness in their totality. In Figure 13 we have an example of the notation used in that literature. Dr. Nader's work establishes connections between each of these 27 branches and specific parts of the human physiology. An example of these links is presented in Figure 14, reprinted from his book.

Dr. Nader's discovery is significant because it establishes a bridge between Maharishi's Vedic Approach to Health and modern medicine. This makes it possible to establish deep conceptual links between the contemporary medical and the Vedic concepts of physiology, health, and disease. These conceptual links provide a sound logic for understanding the preventive and therapeutic methods of this approach and how they successfully address the most challenging problems of contemporary medicine, including cancer, heart disease, drug abuse, anxiety, depression, violent behavior, and many others.

Loss of connection with the unified field of natural law: the basic cause of disease

The preventive and therapeutic methods of Maharishi's Vedic Approach to Health are based on an understanding of the basic cause of disease that is very different from that of modern medicine. According to this approach, disease is caused by violation of natural law. As we discussed above, the unified field of natural law gives rise to a whole array of specific, localized laws of nature that act within the physiology to guide proper, healthy functioning. These localized laws generate a style of physiological functioning that supports health. But two other factors influence

MAHARISHI'S VEDIC ENGINEERING—GENETIC ENGINEERING

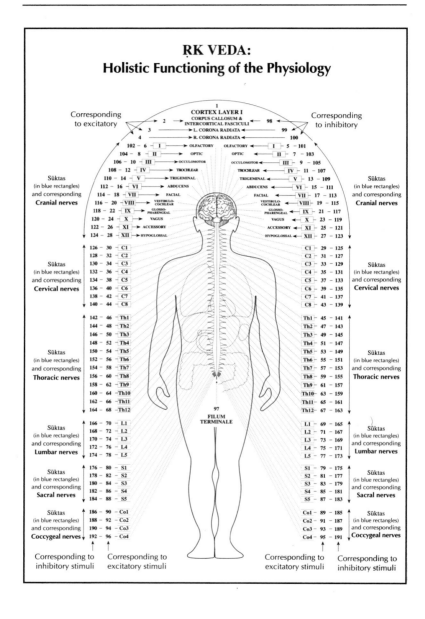

the functioning of the physiology: our actions and the environment. Every day, through our actions, we subject our body to external influences—such as diet, daily routine, stress and strain of daily living, and environmental pollution—which can oppose the healthy dynamics that the laws of nature naturally create within the physiology. This "opposition" amounts to violation of the laws of nature and can trigger new dynamics, a new style of functioning that leads to disease and ill health.

These dynamics first manifest as temporary imbalances in the physiology. However, if they persist long-term, the physiology in effect loses its memory of what the natural healthy style of functioning should be. Rigid blocks of stress accumulate that do not disappear simply by correcting the behavior that created the initial imbalance.

For instance, a combination of improper diet, inadequate exercise, and high stress levels will cause imbalances that lead to the accumulation of plaque in the coronary arteries, and eventually to occlusion of those arteries and heart attack. Removing the stresses that led to this imbalance may prevent the condition from worsening, but the plaque and other impediments disallow the physiology from bouncing back to normal healthy functioning.

Figure 14. Expression of Veda and the Vedic Literature in the Human Physiology (page 86).
This chart presents an overview of the 192 suktas, or verses, of the 1st Mandala, or book, of Rik Veda as they correlate with the nervous system and the entire physiology: Sukta 1 corresponds to layer 1 of the cerebral cortex; complementary Sukta 97 to the silent filum terminale; Suktas 2-4 and Suktas 98-100 to excitatory and inhibitory stimuli of the corpus callosum and corona radiata; Suktas 5-28 and 101-124 to excitatory and inhibitory stimuli of the 24 cranial nerves; Suktas 29-96 and 125-192 to the excitatory and inhibitory stimuli of the spinal nerves.

The central strategy of Maharishi's Vedic Approach to Health

Two things are needed to restore optimal health. First, our actions must be brought into harmony with natural law, so that we do not oppose the dynamics of those laws that support our health. Second, the memory of healthy functioning must be restored to the physiology, and rigid blocks of stress must be dissolved.

Modern medicine attempts to accomplish this piecemeal by manipulating isolated laws of nature, such as physical removal of plaque from the arteries of patients with cardiovascular disease. However, this approach fails to reliably restore health because these isolated laws are innumerable, and the symptoms treated are only the endpoints of complex disease processes.

Maharishi's Vedic Approach to Health uses a fundamentally different and far more practical and effective strategy. Because all of the individual laws that govern the functioning of the physiology emerge from the unified field of nature's intelligence, this approach works on the level of this field to globally restore balanced and integrated functioning on every level of life—from our DNA, to our cells, to our organs and whole body, and to the environment.

Maharishi's Transcendental Meditation and TM-Sidhi program: the keystone of Maharishi's Vedic Approach to Health

Maharishi's Vedic Approach to Health is a coherent, integrated knowledge-based system that makes use of more than 36 distinct modalities to enliven the inner intelligence of the body so that it spontaneously maintains health, and heals itself. Maharishi's Transcendental Meditation and TM-Sidhi program is primary among these modalities. In one stroke this powerful program accomplishes

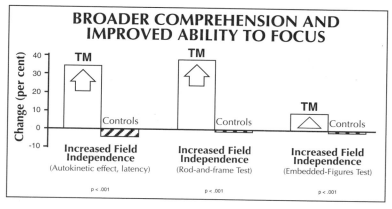

Figure 15. Transcendental Meditation Produces Broader Comprehension and Improved Ability to Focus
Field independence has been associated with a greater ability to assimilate and structure experience, greater organization of the mind and cognitive clarity, improved memory, greater creative expression, and a stable internal frame of reference. This chart shows that practice of Maharishi's Transcendental Meditation technique develops greater field independence. It is notable that no other technique has been found that can systematically develop field independence. Previous to this research, it was thought that this characteristic did not improve after one reached adulthood. *Perceptual and Motor Skills*, 39 (1974): 1031–1034.

---

both requirements for restoring health: it brings our actions into harmony with natural law and restores the memory of healthy physiological functioning.

As described earlier, the Transcendental Meditation technique is a simple, natural technique that allows the mind to gain a unique state of restful alertness, or pure consciousness. Therefore, every time one practices the Transcendental Meditation technique, one gains direct experience of the unified field of natural law. This experience enlivens the qualities of this field in the mind, so that one's thoughts, and consequently one's actions, come into accord with this unified field of intelligence.

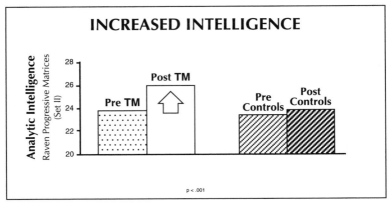

Figure 16. Increased Intelligence through Transcendental Meditation
During a two-year period, university students who practiced the Transcendental Meditation technique increased significantly in intelligence, compared to matched control subjects. Since IQ usually does not change once one reaches early adulthood, this finding reveals unique capabilities of Transcendental Meditation. This result has been corroborated by two independent studies.
Reference: *Personality and Individual Differences*, 12(1991): 1105–1116.

In practical terms, this means that Transcendental Meditation enables individuals to think more clearly, to have broader comprehension, and to approach whatever they do with more creativity. This is experienced on a daily basis by those who practice this technique, and their personal experience has been confirmed by extensive scientific research, some of which is presented in Figures 15, 16, and 17. On the basis of this evidence, it is clear that the Transcendental Meditation technique accomplishes the first requirement for restoring health: it brings our actions into accord with natural law.

The benefits of Transcendental Meditation are profoundly amplified by the more advanced TM-Sidhi program, which includes Yogic Flying—the most powerful of all techniques for enhancing mind-body coordination and enlivening the unified level of nature's intelligence in individual and collective life.

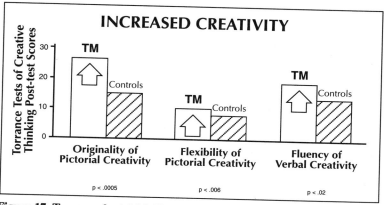

Figure 17. Transcendental Meditation Enlivens Creativity
Individuals who practice Maharishi's Transcendental Meditation technique increased in creativity over a five-month period, compared to matched controls. The Torrance Tests of Creative Thinking showed that verbal fluency, as well as pictorial originality and flexibility, increased. *The Journal of Creative Behavior* 13 (1979): 169–180

Practice of the Transcendental Meditation and TM-Sidhi program also accomplishes the second requirement for restoring health. It restores to our cells and to the physiology as a whole the "memory" of healthy functioning.

The mind and body are intimately connected. As the mind settles down during the practice of Transcendental Meditation, the body also settles down to a deep state of rest. Therefore, when the mind attains the state of restful alertness and expanded consciousness, the functioning of the body is transformed as well. As shown earlier in Figure 11, several physiological measurements indicate that Transcendental Meditation provides very deep rest. This rest gives the body an opportunity to repair the physiological blocks and impediments that are associated with disease.

In addition, when the mind identifies with the unified field of nature's intelligence, the physiology restores its connection with the full value of that intelligence. This re-enlivens in the physiology the memory of what proper, healthy functioning really is, so that each molecule in each cell of every tissue and organ of the body moves toward a style of functioning that supports health.

As we discussed earlier, extensive research has been done on Maharishi's Vedic Approach to Health. The Transcendental Meditation and TM-Sidhi program is the most thoroughly studied of the modalities of this system. When this body of research is compared to that of any modern medical treatment or drug, one thing is strikingly absent: that is, the lack of negative side effects. In the over 4,000 pages of research papers compiled in *Scientific Research on the Transcendental Meditation and TM-Sidhi Program: Collected Papers Volumes 1-6*, it is notable that no harmful effects are observed. This confirms the point made earlier that these technologies, which work from the level of the unified field of natural law, the field of pure consciousness, create integrated beneficial effects—and no harmful side effects—in all aspects of the mind and body.

Technologies that work from this fundamental level of natural law are not only safer, but also more effective than technologies that work from more superficial levels. By working at this fundamental level, it is possible to eliminate those imbalances that are the source of disease, instead of waiting for disease to manifest fully and then treating the symptoms.

Restoring healthy functioning of the physiology through the other modalities of Maharishi's Vedic Approach to Health

Practice of Maharishi's Transcendental Meditation and TM-Sidhi program immediately begins to correct the basis for disease, by restoring the individual's connection with the unified field of natural law, and reducing stress and disturbances in the physiology. However, with long-standing, chronic disorders, such as coronary artery disease, the other 35 modalities of Maharishi's Vedic Approach to Health serve an important function. These include diet, nutrition, exercise, herbal supplements, and procedures for physiological purification and detoxification. These are all designed to accelerate the removal of abnormalities and disturbances, and to accelerate the process of restoring the body's memory of correct physiological functioning.

The central strategy of each of these modalities is to provide knowledge or intelligence that restores balanced and integrated functioning to the mind and body. When the mind and body achieve this state of inner balance and harmony with the laws of nature, good health is the spontaneous consequence. These numerous modalities work from the angles of consciousness, breath, body, behavior, and environment to enliven nature's intelligence, and thereby restore, preserve, and promote health.

Maharishi Ayur-Veda Universities

Maharishi's Vedic Approach to Health is available worldwide through Maharishi Ayur-Veda Universities. To date, 22 Maharishi Ayur-Veda Universities have been established in major cities throughout the U.S., and another 28 are planned, with the goal of founding one in every state. Similar efforts are in progress in Europe, Asia, Africa, South America, and Australia.

These institutions serve as educational, research, and treatment centers. They offer degrees in Maharishi's Vedic Approach to Health and certification programs in which experienced physicians learn to incorporate this approach into their own practices.

Already more than 300 physicians in the U.S. have been trained in Maharishi's Vedic Approach to Health, and several thousand more have been trained worldwide. In addition to using this approach in the context of their own medical practices, physicians have established clinics devoted to this approach in large metropolitan areas, not only in North America, but also in Europe, Australia, South America, and Africa.

Maharishi Ayur-Veda Universities offer educational programs to the general public on diet, nutrition, exercise, and many other topics. The objective of these programs is to assist individuals in developing systematic prevention strategies for themselves, so that they can be self-sufficient in maintaining health and treating disease.

Among the other programs offered by these universities are intensive programs designed to bring relief from a wide range of chronic diseases, including coronary artery disease, bronchial asthma, hypertension, irritable bowel syndrome, atherosclerosis, insomnia, arthritis, and migraine headache.

Enlivening collective consciousness

One absolutely critical question remains: How do we ensure that physicians, patients, and biomedical researchers adopt safe, natural preventive approaches and give up their dreams of magic bullet, high-tech approaches, including hazardous gene therapies?

Knowledge regarding the dangers of genetic engineering and knowledge of the benefits of effective, natural preventive approaches, such as that presented in this book, will have some positive impact. However, knowledge must be *applied* to be of benefit. In order to ensure that this knowledge is used, it is essential to maximize the clarity of thinking and depth of comprehension of all those who participate in decisions regarding health care strategies.

These decisions are made by physicians, scientists, businessmen, law-makers, and the public at large. The quality of these decisions is determined by how awake, aware, and alert—how conscious—these individuals are. This, in turn, is profoundly influenced by the level of stress in society and the quality of collective consciousness.

The inadequacies of our medical system can ultimately be traced to the current status of collective consciousness. Because awareness has been restricted to surface levels, thinking has not been sufficiently comprehensive and coherent. Consequently, patients, physicians, and researchers see only the parts of human life, and have lost sight of the fact that human life is an integrated whole. Physicians have lost sight of the fact that health cannot be maintained or restored by dealing with isolated parts, and have focused on progressively more isolated values of the physiology in their attempts to preserve and restore health. The move to dangerous and destructive genetic therapies is the logical and most extreme outgrowth of this trend. To reverse this trend it is essential that collective consciousness be enlivened.

The Maharishi Effect

Numerous scientific studies have established that the public health modalities of Maharishi's Vedic Approach to Health, most importantly the group practice of Maharishi's Transcendental Meditation and TM-Sidhi program, are very effective tools for enlivening collective consciousness.

To date, 42 research studies have provided strong scientific evidence that crime, sickness, and accident rates all decrease highly significantly when one percent of a population practices Transcendental Meditation or when the square root of one percent of the population regularly practices the Transcendental Meditation

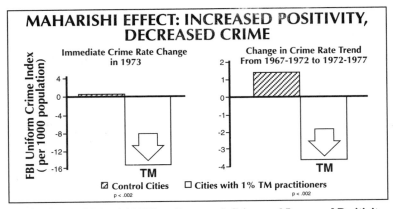

Figure 18. The Maharishi Effect: Decreased Crime and Increased Positivity
Twenty-four cities in which 1% of the population had been instructed in the Transcendental Meditation program by 1972 displayed significant decreases in crime rate during the next year (1973), and displayed decreased crime rate trends during the subsequent five years in comparison to the period from 1967-1972. In contrast, crime increased in control cities, matched for geographic region, population, socio-economic and educational distribution of the population, and initial crime rate. *Journal of Crime and Justice*, 4 (1981): 25–45.

---

and TM-Sidhi program as a group. This positive social influence has been termed the Maharishi Effect. Figures 18 and 19 present examples of research done on this effect, showing the power of these Vedic technologies in reducing crime. Research on the Maharishi Effect has been published in leading peer-reviewed journals, including the *Journal of Conflict Resolution, Social Indicators Research,* and *Mind and Behavior* (28, 29, 30, 31). The fact that so many studies report similar results shows that this effect is highly reproducible. Furthermore, the 42 published studies show that the Maharishi Effect can be induced in populations of diverse social and cultural origins, in many areas of the world. Thus, it is broadly applicable.

These studies also show that implementing the group practice of Maharishi's Transcendental Meditation and TM-Sidhi program is feasible in practical terms. To create beneficial effects, it is not necessary for everyone to practice this program: it requires only a small portion of the population—the square root of one percent. A group of fewer than 2,200 individuals would be required for Europe or North America. Only about 7,000 would be required for the world as a whole. Considering that the standing army of even the smallest nation exceeds these numbers manyfold, it is within the budgetary capabilities of any nation to establish such a group.

Reduction in crime, sickness, and accident rates indicates broader comprehension and improved decision-making among the whole population. These reductions indicate that people are making better choices, resulting in safer, more healthful behavior; thus they avoid accidents and illness. This technology seems to influence the ability of even the most stressed individuals in society to make better choices, since crime—including violent crime—diminishes as well (see Figures 18 and 19).

Because the quality of comprehension and clarity of thinking are determined by how awake, how conscious an individual is, these reproducible improvements in crime, sickness, and accident statistics indicate that group practice of Maharishi's Transcendental Meditation and TM-Sidhi program is a powerful tool for reducing collective stress in society and for developing collective consciousness.

How does this relate to genetic engineering and health care? If this tool can be used to improve clarity of thinking of the members of society, thereby reducing crime, sickness, and accidents, it should also improve the clarity of thinking and decision-making ability of those—ultimately all of us—who determine how society approaches health problems. This technology can promote a less stressed, more awake, more conscious environment in which

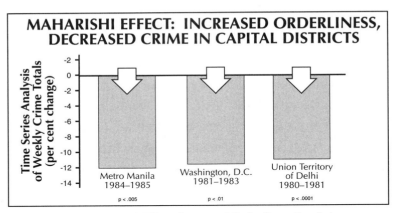

Figure 19. The Maharishi Effect: Increased Orderliness Leads to Decreased Crime in Three Capital Cities
Crime decreased significantly during periods in which the number of group participants in Maharishi's Transcendental Meditation and TM-Sidhi Program exceeded the square root of 1% of the population of these three capital regions—Metro Manila, Philippines (August 1984–January 1985); Washington, D.C., U.S.A. (October 1981–October 1983); Union Territory of Delhi, India (November 1980–March 1981). *Journal of Mind and Behavior* 8 (1987): 67–104 and 9 (1988): 457–485.

---

everyone—physicians, scientists, businessmen, law-makers, and the public—is able to make wiser, more far-sighted decisions on all health issues, including those related to genetic technologies.

Because group practice of Maharishi's Transcendental Meditation and TM-Sidhi program is capable of accomplishing this aim, and is the only scientifically validated method available, this technology—a central aspect of Maharishi's Vedic Engineering—fulfills a unique role in protecting humanity from the hazards of genetic engineering. Without this technology for developing collective consciousness, the broader comprehension will not be available that is required to make the choices needed to transform the currently destructive direction in medicine into one that delivers health.

Maharishi's Vedic Approach to Health: a foundation for effective medical practice

Modern medicine has given us details of physiological structure and function, but what has been missing is an understanding of the intelligence which underlies, unifies, and structures the physiology. Without this understanding, medicine has been without a basis.

Maharishi's Vedic Approach to Health, which is central to Maharishi's Vedic Engineering, provides that basis. This powerful knowledge-based approach provides knowledge of the unified field of natural law, knowledge of the physiology, and knowledge of the relationship between these two. From this comes applied knowledge of a whole range of technologies that awakens nature's inner intelligence within the mind and body, thereby enlivening the body's own healing powers.

These technologies provide the physician and patient with systematic, effective methods for restoring and maintaining balance in the physiology and thereby efficiently fighting disease and maintaining health. Through this strategy, it is possible to address a wide range of health problems that modern medicine has previously been powerless to solve.

PART II
AGRICULTURE

CHAPTER 4

THE HARVEST OF AGRICULTURAL GENETIC ENGINEERING: GENETIC POLLUTION AND DISRUPTION OF THE ENVIRONMENT

Today, virtually every major player in the agricultural industry has joined the race to derive commercial gains from genetic technologies. To the extent that agricultural applications involve the release, purposeful or accidental, of genetically engineered organisms into the environment, they pose a serious risk. At this time, this risk is imminent because the implementation of these technologies is being guided, not by scientific evidence, but by economic and political imperatives.

Extensive environmental release

The extensive release of genetically altered agricultural varieties is already taking place. From 1987 to the spring of 1994, over 2037 small-scale releases occurred in the U.S. alone, as part of field tests. The number of releases almost doubled during the first half of 1994, reflecting the rapidly accelerating rate of research in agricultural genetic engineering. Research-scale releases are also common in many other countries. Even these test releases carry some risk. However, commercial use of genetically engineered organisms in agriculture is much more dangerous because it will greatly widen the scope of release, in some cases to millions of acres. Such releases are just beginning. Already, two genetically altered tomato varieties have been cultivated on a large scale in the U.S., and a squash has recently been cleared by the Environmental Protection Agency (EPA) and Food and Drug Administration (FDA) for commercialization.

Lax regulations

The risks associated with the accelerating rate of research and development in agricultural genetic engineering is exacerbated by the lack of stringent oversight by regulatory agencies. Regulation is weak or non-existent in almost all countries. For instance, the U.S. regulatory system has substantial and obvious weaknesses. As they stand, regulations regarding genetic manipulations of plants, animals, and microorganisms are a hodgepodge that is administered by three different agencies, the U.S. Department of Agriculture (USDA), the EPA, and the FDA (32). Lines of jurisdiction between these three are blurred in some areas, and there are loopholes and large gaps in regulation. For example, the release of genetically engineered aquatic species is not covered under any of the existing regulations. Thus at this time, such releases could be carried out by anyone without legal repercussions. This is a situation in which serious mistakes could occur. Concerned environmental scientists are calling for improvements in regulations (33), but progress is glacially slow. Jurisdiction over release of genetically engineered microorganisms is also questionable. These are currently regulated by the EPA under the Toxic Substances Control Act, which was designed to control toxic chemicals, not genetically engineered microorganisms (32). The inclusion of genetically engineered organisms under this act is tenuous and has yet to be tested in the courts.

In addition to weak regulations, political and economic pressures often override safety issues. The biotechnology industry has repeatedly and unfairly blamed its slow growth on the inhibitory effects of regulatory agencies. As a result politicians, concerned about votes, have put considerable pressure on regulatory agencies to facilitate the review process. These pressures have been ongoing for many years and have therefore significantly altered the mindset within these agencies. Personnel have realized that professional rewards will come, not from protecting the public from potentially dangerous technologies, but from speeding movement of technologies through the review process. The result is that the effectiveness of these agencies in safety assessment has been greatly weakened.

Hazards of environmental release of genetically engineered organisms for agricultural purposes

As summarized in Chapter 1, the main hazards of agricultural genetic engineering are as follows:

- disruption of the ecosystem, either locally or globally, leading to

loss of biodiversity and disruption of the food chain, with both environmental and economic impacts, including:
- reduction of soil fertility,
- weakening or destruction of species that are environmentally or commercially important,
- destruction of centers of biodiversity for important food crops;

- increased use of carcinogenic and mutagenic agricultural chemicals, leading to water pollution and in turn to increased incidence of cancer, birth defects, and other illnesses;

- creation of new plant diseases, pests, and weed varieties that are resistant to known antibiotics, pesticides, and herbicides.

Each of these dangers is illustrated in the examples presented in the following sections. Impartial analysis reveals that one or more of these hazards is associated with virtually every one of the hundreds of other genetically engineered organisms that have been developed and are currently under review by regulatory agencies.

Crops genetically engineered for herbicide resistance

Herbicide-resistant crops would appear to provide a convenient strategy for weed control. They make it possible for the farmer to spray the fields liberally with herbicide. This kills the weeds, but the crop plants, which are resistant to the herbicide, survive. However there are serious disadvantages to this strategy, and when these are included in the equation it becomes clear that the only gains from this approach are the economic rewards to the chemical companies who develop these crops (34, 35).

A strategy for increasing sales of agricultural chemicals

Conservative calculations indicate that widespread cultivation of herbicide-resistant crops will increase herbicide use at least three-fold. Thus, use of herbicide-resistant crops will strongly boost herbicide sales. Not surprisingly, each of the major agricultural chemical companies has been very actively developing crops resistant to their own herbicides. These companies include Monsanto, DuPont, Calgene, American Cyanamid, Rhone-Poulenc, and many others. Their efforts encompass essentially every major field crop grown around the world, including corn, soybean, rice, cotton, potato, wheat, and sugar beet. Other important crops have also been engineered for herbicide resistance, including tomato, carrot, cucumber, eggplant, lettuce, pea, pepper, squash, apple, cantaloupe, melon, papaya, peanut, plum, sunflower, walnut, and watermelon.

If this were a benevolent scheme for increasing herbicide sales, it might be tolerable; but that is not the case. Agricultural chemicals are already the primary contributors to water pollution. These genetically engineered crops will bring at least a three-fold increase in herbicide use, making the pollution problem much worse. Eighty percent of ground water in the U.S. is already contaminated by these chemicals. Similar problems exist in Europe and other areas of the world. In some areas of the U.S., by the age of six, a child will have consumed in drinking-water 130% of the life-time recommended safe limit of herbicides. Some herbicides, such as atrazine, are carcinogens while others, such as bromoxynil, cause birth defects. Epidemiological studies have identified herbicide contamination as the cause of increased rates of cancer in certain regions of the U.S.

Herbicide-resistant weeds

On top of these harmful effects on health, genetically engineered herbicide resistance does not even offer a definitive solution to the weed problem. Herbicide-resistant crops are only a temporary and partial solution. The genes for herbicide resistance will flow into related wild species via cross-pollination, generating in only a few growing seasons weeds that are resistant to the herbicide. In addition, it is well known that weeds spontaneously develop resistance to herbicides. In the long run, development of herbicide-resistant crops will create greater weed problems for the farmer.

A clear case of "bait and switch"

The development of herbicide-resistant crops is a clear example of industry saying one thing and doing another—the "bait and switch" routine. Initially, industry justified going into genetic engineering by claiming that it would lead to a kinder, gentler approach in agriculture, one that would reduce dependence on chemical herbicides and pesticides, and therefore reduce chemical impacts on the environment.

In contrast to this stated intention, the development of herbicide-resistant crops will have just the opposite effect. It will drastically increase herbicide use. Development of herbicide-resistant crops has been the predominant thrust in industrial genetic engineering research and development—47% of all genetically engineered crop plants developed to date carry an herbicide-resistance gene. Industry has invested as much in this one area as it has in all other agricultural applications of genetic engineering combined. Thus, industry's primary investment has been in applications of genetic engineering that harm the environment, not in ones that might benefit it. Obviously, industry's first and primary interest in genetic engineering has been to implement strategies for increasing profits, not ones that improve agricultural practices or protect the environment.

Making a profit wins out over humanitarian concerns

In addition to engineering crops resistant to more modern, "environmentally friendly" herbicides, agricultural chemical companies have genetically engineered crops resistant to older herbicides which are stable in the soil and more toxic, such as 2,4,-D and atrazine, both of which are carcinogenic. Use of these crop varieties will not be widespread in the U.S. or in other developed nations. The chemical companies will sell these more toxic herbicides in nations where safety standards are lower.

In those countries, herbicide application is likely to be accomplished by hand. Therefore, the exposure of farm workers to these carcinogenic materials will be greatly increased through the use of genetically engineered crops and the resultant increase in chemical use. It is obvious that the profit motive vastly out-weighs humanitarian concerns in those who are developing these genetically engineered crops.

Genetically engineered bacteria can disrupt soil ecology

We tend to think of applications of genetic engineering in microorganisms as the safest, since these are the simplest of all organisms. However the following example shows that when the application involves release into the environment, the exact opposite is true.

Alcohol-producing Klebsiella

Klebsiella planticola, a bacterium that normally lives in the soil, has been engineered to help dispose of agricultural and lumber industry wastes—corn stalks and wood chips—and to generate ethanol for fuel. The initial plan was that, once farmers had used this organism to convert waste biomass into ethanol, they could spread the residue from this process on their fields, like compost. The problem with this plan is that this residue is toxic to soil. When planted in soil mixed with this residue, seeds sprout, but then wilt and die (36, 37).

It turns out that this genetically engineered bacterium is highly competitive with native soil microorganisms. Since the bacterium persists long-term, its influence is likely to be long-term. The *Klebsiella* take over, strongly suppressing microorganisms that are important for soil fertility. For instance, the genetically engineered *Klebsiella* significantly reduce the mycorrhizal fungi that most plants require in order to take up nitrogen and other nutrients from the soil. These genetically engineered bacteria also drastically change other components of the community of soil microorganisms and bring about other complex changes in soil chemistry. This, and the fact that these bacteria also produce ethanol—which is toxic to plants and to some microorganisms—partially explains their deleterious effects on plant growth.

Based on these observations, the environmentally responsible action would be to discard this genetically engineered microorganism and return to the drawing board. However, economic and political pressures are moving things in a very different direction. In the name of spurring the economy, there is strong pressure from both industry and government to allow large-scale use of this bacterium.

Unpredictable effects of bacteria introduced into the soil

Soil microbiologists and ecologists assert that we do not know nearly enough about soil microbiology to be able to predict the effects of introducing hardy strains of foreign bacteria into the soil. They estimate that there are more than 1600 different species of microorganisms in a single teaspoon of soil. At best, only 10% of those species have been named, not to mention characterized. It is virtually impossible to understand the possible interactions among all of those species well enough to be able to model or predict the effects of adding a new species to the soil.

Until this complex area of ecology is much better understood, it would be foolhardy to release genetically engineered bacteria into the soil on even a moderate scale. Experts in this area assert that it will be necessary to carry out extensive and systematic empirical research with each microorganism when its release is being considered, in order to assess, even crudely, the potential environmental effects of a specific release. Even then there will be surprises. Given the complexity of the system, it is unlikely that it will ever be possible to model the effects of such releases with long-term accuracy and safety.

Genetically engineered livestock

Many wild or semi-domesticated species are now being genetically manipulated to optimize traits of commercial interest. These organisms are intended for culture in captivity. However, because they are capable of living in the wild, it is inevitable that a few will escape, and with them, altered genes will enter the gene pool of the wild population. This can have serious effects on the wild population and on the ecosystem (33). Most critically, it will be

GENETIC POLLUTION AND DISRUPTION OF THE ENVIRONMENT

impossible to control or eradicate a genetically engineered organism if it causes problems once it has been released into the wild. Thus effects on the ecosystem are irreversible.

Salmon, for example, are being engineered to have an extra growth hormone gene (38). As a result, the salmon grow much larger, but with considerable variation—3 to 15 times normal size. Several problems could arise as a consequence of accidental release of these fish. The two most critical are outlined in the following sections.

Genetic pollution

Due to the limitations of the genetic engineering methods of today, these genetically altered salmon, especially the larger ones, often have developmental defects. If these salmon escape, the genes responsible for these defects will enter the salmon gene pool, which could weaken the whole species. Thus in the process of attempting to improve specific traits through genetic manipulations, we may accidentally disrupt some other important trait, thereby reducing fitness.

Disruption of the food chain and the ecosystem

In the past, ecosystems have been seriously disrupted by the introduction of organisms native to some other area of the world. When we genetically engineer an organism we are, in effect, creating a new species. When such novel organisms are released into the environment, they could be even more disruptive than transplanted natural organisms. For instance, dramatically changing the size of an organism, as has been done with salmon, could disrupt the food chain quite significantly—what used to be prey may now become predator. Alterations in other traits, such as environmental tolerances (*e.g.*, temperature or salinity tolerance), behavior (*e.g.*, migration, mating, territoriality), metabolic rate, and disease resistance, would also substantially influence the interactions of genetically altered fish with the ecosystem.

Genetically engineered squash

The FDA and USDA have recently given permission to the Upjohn Corporation to commercialize a squash that has been genetically engineered to be resistant to a virus. Permission was granted even though scientists and public interest groups raised strong arguments, backed by scientific evidence, that the FDA and USDA safety analyses of this squash were flawed and scientifically inadequate.

Although this genetic engineering approach may help the farmer to deal temporarily with one particular virus, viruses and other pathogens are known to change relatively quickly in order to circumvent plant resistance mechanisms. In a few years, the resistance trait that genetic engineers have so laboriously built into this squash will be obsolete. Moreover, the genetic alterations carried out do not confer protection from any of the other viruses or other pathogens with which the farmer must contend. Such protection would require introducing a resistance gene for each pathogen or pest, which is not feasible. In addition, this genetically engineered plant poses several hazards (39), the most serious of which are summarized below.

Destruction of biodiversity

Genes that confer resistance to a virus could flow from the squash to wild relatives. This resistance gene would make these squash relatives more competitive with other wild species. The engineered squash itself could also move out of the field and compete with other plants. As a result, either the virus-resistant squash or virus-resistant relatives could take over niches in the ecosystem that other plants naturally occupy.

This displacement could reduce the biodiversity of that ecosystem. The members of an ecosystem are interdependent in highly complex ways. When one species is lost, other members of that ecosystem that are dependent on it will suffer and may themselves be unable to survive. For instance, the plant initially displaced from the ecosystem might be required as food by one species and might be required in the reproduction of another species. When that plant is excluded from the ecosystem, the two dependent species may be lost as well.

Likewise, survival of other species could be dependent on those two, and would be threatened by their displacement. This domino effect can amplify an initial, minor alteration in the ecosystem, causing much more substantial effects over time. Due to the complexity of the ecosystem and to the potential for highly complex interactions, possible side effects can emerge after many years. These side effects would have been totally unpredictable at the time the initial genetic change was introduced.

Loss of biodiversity not only damages the ecosystem but can also be economically harmful. In particular, it can destroy the centers of biodiversity for important food crops. For instance, teosinte—the wild plant from which corn, or maize, is derived—still grows wild in Mexico and other areas of Central America. The center of biodiversity for potatoes resides in wild relatives in South America, while that for squash resides in wild plants in North America. When a new pathogen or crop pest appears, plant breeders go back to these wild repositories of squash or teosinte or potato genes to breed into commercial strains new traits that will increase resistance to that pathogen or pest.

Reducing biodiversity in these wild populations will diminish the genetic resources that traditional plant breeders have to work with. This will make the development of future crop plant varieties even more dependent on genetic engineering because the resources for traditional breeding will be unavailable. This would be very unfortunate because genetic engineers are unlikely to be able to match nature's own creativity in the development of useful new traits.

More troublesome weeds

Genes that confer resistance to a virus could flow from the squash to wild relatives, making them more hardy and, therefore, weedier weeds. The engineered squash, itself, could also behave as a weed.

Generation of new viruses

Recent research has revealed an even more serious problem (40, 41). Because of the way in which this squash was genetically modified to be resistant to one kind of virus, there is a significant risk that when this plant is infected with other viruses, genetic events may occur that would generate new viral strains—and thus new crop diseases. In effect, these virus-resistant plants could serve as an incubation chamber for generating new viruses that are capable of infecting a wider range of plants. A genetic alteration designed to help farmers could, in the long run, create much greater problems for them.

Genetically engineered bovine growth hormone

The examples described above involve the release of genetically engineered organisms into the environment. There are also applications of genetic engineering in agriculture that do not involve environmental release. The use of bovine growth hormone (BGH) to stimulate milk production in cows is an example of such an application. Based on early research demonstrating that BGH stimulates milk production in cows, Monsanto, Eli Lilly, Upjohn, American Cyanamid, and a number of other companies have genetically engineered bacteria to produce large amounts of this hormone. These companies have invested over $1.1 billion in genetically engineering the production process for BGH and in testing it for use in dairy herds.

Sick cows yield inferior milk

In carrying out further research and development, these companies and independent investigators learned that, although BGH increases milk production by as much as 15%, it is bad for cows. It has a number of harmful side effects, including clinical mastitis, cystic ovaries, reduced pregnancy rates, and other reproductive problems. These translate into increased veterinary costs and increased use of antibiotics and other drugs. While the milk from these cows has not been found to be overtly toxic, milk produced by sick cows is less healthful. It is inferior because mastitis results in secretion of increased amounts of white blood cells (also called "pus") into the milk (42), and because increased use of antibiotics and other drugs leads to residues of these drugs in the milk (42).

It is clear from the research literature that these side effects are sufficiently severe that, in all good conscience, these companies should have ceased development efforts and withdrawn plans to market this product. However, the companies involved, especially Monsanto, have taken quite the opposite direction: they have pressed forward vigorously with commercialization. According to an article in the leading scientific journal, *Nature*, at least one of these companies has suppressed data regarding negative side effects (42). Other sources report that the regulatory agencies responsible for evaluating the safety and efficacy of this product have been manipulated in order to facilitate approval (43, 44). These companies have also mounted an aggressive sales campaign aimed at convincing farmers that use of BGH will be good for their business.

Clearly the objective of these actions is, above all else, to recoup initial investments and to make profits. These actions demonstrate that—in the current social climate—when economic and commercial priorities come into conflict with health and humanitarian considerations, it is economics that will prevail. These companies have clearly put profits above the quality of the milk that we feed to our children, and above the health, well-being, and comfort of millions of dairy cows. To many it would seem cruel to treat a cow with a drug that makes it sick and uncomfortable. What ever happened to the idea of wholesome milk from contented cows?

Synergy between over-optimistic scientists and over-eager businessmen

The marketing of BGH to the dairy industry illustrates the dynamics through which economic priorities overshadow health and humanitarian priorities in corporate research and development. It also illustrates another point relevant to commercialization of genetic technologies. The development and application of BGH show how synergism, between scientists who are over-enthusiastic about genetic technologies and businessmen who are over-eager to profit from those technologies, can create products that no one really wants and that decrease the quality of life for everyone. After all, there is already a milk surplus in the U.S. and in Europe. What is the logic of sacrificing milk quality to increase production further?

Scientists over-promote the commercial potential of their basic research discoveries for three reasons. First, scientists have a natural enthusiasm for what they are studying in the lab; second, they are generally naive regarding the real challenges of translating a lab-bench finding into a product ready for market; and third, they are eager to follow in the steps of virtually all of the leaders in the field of molecular biology, who have become wealthy by translating their basic research into profitable commercial enterprises. The business community has tended to over-respond to this enthusiasm—and to the credentials of these scientists—investing too much money too quickly, and only later encountering unanticipated problems in commercialization. When problems arise, the tendency is to press forward to recoup investments, despite negative impacts on society. This line of action is neither ethical, nor, in the long run, profitable.

This syndrome has already significantly weakened the genetic engineering industry in two ways. First, rushing ahead with commercialization has resulted in substantial losses when products fail late in development due to unanticipated side effects. According to one expert in industry, "Biotech companies do the minimum early development, leaving themselves open for later shocks."(45) Second, the frequency with which such failures have occurred has sharply reduced the confidence of investors. As a result the industry is in a precarious condition. Of approximately 250 companies listed on the stock exchange, 98—or about 40%—experienced a drop of 50% or more in stock value during 1994 (45).

As interest from smaller investors has evaporated, the multinational pharmaceutical companies have moved in, seeing this as an opportunity to buy into these companies at bargain-basement prices. Given the past record of the multinationals, their growing influence in the biotech sector is not likely to promote greater emphasis on safety.

Partial solutions and novelty products

The examples presented above illustrate a number of significant problems that can arise in using genetic engineering in agriculture. The natural question to ask is whether these products contribute sufficiently to agricultural practice or production to warrant the associated risks. The following table summarizes the percentage of field trials for different traits related to the total number of field trials for genetically engineered crops carried out from 1985 to 1994 in the U.S. It shows that five different classes of traits are being engineered into crop plants (46).

Trait	Percentage*
Herbicide tolerance	47
Insect resistance	25
Altered crop quality	20
Virus resistance	17
Bacterial and fungal resistance	5

* Percentages total more than 100 because multiple traits were engineered into some plants.

We saw above that herbicide resistance provides only short-term, partial solutions to weed problems, because of gene flow and because weeds adapt to herbicides. Likewise, we saw that genetically engineered virus resistance is not only circumvented by viral adaptation, but that the interaction of the genetically engineered plant with other viruses may actually generate new plant viruses, new diseases. Similarly, genetically engineered insect resistance and resistance to bacterial and fungal infections are temporary solutions because of adaptation to pesticides and antibiotics.

To perpetuate this approach, biotech companies will have to continually develop new genetically engineered varieties to keep up with pathogens' natural adaptive abilities. From the perspective of the agricultural multinationals, this is ideal. Programmed obsolescence will always ensure a market for their newest products. However, it is far from ideal for the farmers—and the consumers—who must ultimately bear the costs of developing those new products. It means increased operating costs for farmers and increased food costs for everyone. This approach also erodes the self-sufficiency of the farmer.

In addition, this approach fails to accomplish the stated goal of agricultural genetic engineering: to eliminate or substantially reduce use of agricultural chemicals. Because it is not practical to engineer more than a few traits into a plant, genetic engineers will never be able to design crops that do not require the use of some classes of agricultural chemicals. Therefore, this approach can only be viewed as an adjunct to standard approaches, and only a partially successful one at that.

The final class of genetically engineered crop plants, "crops altered in quality," is equally unnecessary. Crops altered in quality are plants like the FlavrSavr® tomato, which has been genetically engineered to have a longer shelf life. Any manipulation that alters the properties of the food product falls into this category. For instance, Pioneer Hybrid has developed a soybean that provides complete protein. The company did this by introducing the gene for a Brazil nut storage protein into the soybean genome. Such alterations may seem convenient, but properties such as balanced dietary protein can almost always be achieved through other simpler and safer means such as eating certain foods in the proper combinations.

Moreover, these alterations can influence food quality or safety in unexpected ways. For instance, Pioneer has ceased development of this soybean because they discovered that the Brazil nut protein in this soybean causes allergic reactions in a reasonably large portion of the population. Any alteration of the composition of a food plant can have similar harmful effects, and those effects may not be discovered until consumers begin to use the product widely. Why subject the public to this risk, when these products do not offer unique solutions to critical needs, but merely represent novelty items or incremental refinements over existing products?

In summary, the examples presented in this chapter show that the potential damage that can result from releasing genetically engineered organisms into the environment is great and, in some cases, that damage will be irreversible. At the same time, the use of genetically engineered organisms in farming offers, at best, partial solutions to agricultural problems. This benefit/risk ratio is clearly not favorable and does not justify the widespread use of this strategy. This is especially the case when, as will be established in the next chapter, more effective and safer approaches are available.

CHAPTER

FEEDING THE WORLD WITHOUT GENETIC ENGINEERING OR CHEMICAL POISONS— MAHARISHI'S VEDIC APPROACH TO SUSTAINABLE AGRICULTURE

Current challenges in agriculture

The application of chemicals to the soil for the purpose of enhancing productivity—chemical agriculture—was pioneered in the 1800s by the chemist Justus von Liebig. He spent a major part of his career developing and promoting this approach. However later in his life he began to appreciate more fully the wisdom or order inherent in nature and recognized that the approach he had developed interfered with that order. He explained, "In my blindness, I believed that a link in the astonishing chain of laws that govern and constantly renew life on the surface of the Earth had been forgotten. It seemed to me that weak and insignificant man had to redress this oversight" (47).

Unfortunately, mainstream agriculture has yet to appreciate the critical point regarding the completeness and self-sufficiency of natural law that von Liebig had realized over one hundred years ago. In fact, today, heavy dependence on agrochemicals is just one dimension of the high-tech, manipulative approach that has increasingly dominated agriculture since his time.

These approaches increase productivity in the short term, but do so at the expense of the environment. Mono-cropping, over-cultivation, over-grazing, and other non-sustainable agricultural methods result in erosion of billions of tons of topsoil each year in the U.S. alone. Furthermore, chemical fertilizers, herbicides, and pesticides pose increasing problems to farmers and endanger their health. These chemicals add to the operating expenses of the farmer, yet are becoming less effective by breeding increasingly resistant strains of weeds and insects. Agricultural chemicals are also a major source of pollution to ground and surface water. Heavy reliance on chemical fertilizers depletes the natural life and fertility of the soil and therefore poses a serious hazard to future generations. As discussed in Chapter 4, the use of genetic engineering in agriculture will further exacerbate these pollution problems and will generate a host of other new problems that are even more serious.

The two primary applications of genetic engineering in agriculture today are the development of herbicide tolerant crops and the development of crops resistant to pests. A third application is to develop plant varieties with modified functional or nutritional properties. As discussed in Chapter 4, this third category consists of novelty products that do not offer uniquely valuable solutions to critical agricultural or nutritional needs. To put it bluntly, they are superfluous. For instance, the FlavrSavr tomato, which is genetically engineered to reduce spoilage, and which is the first such product to reach the market, will have little fundamental impact on humanity.

The other two categories of agricultural genetic engineering products also turn out to be unnecessary. Through the judicious use of sustainable agricultural strategies, such as crop rotation, crop diversification, and natural pest control methods, the objectives that scientists hope to achieve through genetic manipulations can be accomplished much more simply, safely, and cost-effectively. In addition, sustainable methods can be implemented immediately. We need not wait for scientists to complete many more years of research and development.

Sustainable agricultural strategies

A wide range of methods are available today that can be used to restructure agriculture as a self-sustaining system that supports health and the environment. The central principle of sustainable agriculture is to produce food and fiber for human needs using methods that can be sustained indefinitely. The key to this approach is to restore, preserve, and enhance soil fertility. This must be accomplished using only renewable resources and using methods that do not impact negatively on the environment.

Scientific research has identified numerous effective alternative farming techniques, including use of fertilizers derived from renewable resources, soil-protective tillage methods, crop diversification, crop rotation, and natural methods of pest control. These methods can produce yields as high as current mono-cropping methods that use primarily chemical approaches (48). Yet sustainable methods are safer and less costly, and do not expend non-renewable resources. The effectiveness of these methods make it unnecessary to resort to the use of agrochemicals, genetic engineering, or other damaging practices, and, because these methods are sustainable over the long term, they do not erode the basis of life for future generations.

Renewable fertilizers

Fertilizers derived from renewable resources, such as manure, agricultural wastes and cover crops, free the farmer from dependence on chemical fertilizers. The manufacture of chemical fertilizers is extremely energy intensive and therefore depletes petroleum and other non-renewable energy resources. Healthy soil contains a rich and complex mixture of organic material, microorganisms, and other small organisms, in addition to minerals. These all contribute to the blend of nutrients found in the soil. Organic fertilizers enhance fertility by enriching this mixture and by contributing to the growth and vitality of soil organisms. In contrast, chemical fertilizers contribute high concentrations of specific nutrients, but disturb the balance of the soil and are actually toxic to many soil organisms. The overall result of using chemical fertilizers is, therefore, to degrade the vitality and fertility of the soil.

This was the heart of von Liebig's realization: knowledge of a few isolated laws of chemistry is not an adequate basis for enhancing soil fertility or agricultural productivity in the long term. By interfering with the complex network of interactions among living organisms and the diversity of organic and mineral constituents of the soil, we inevitably do far more harm than good. This conclusion is also obvious from the principles of Maharishi's Vedic Science, discussed in Chapter 3. Fortunately, through application of the holistic and sustainable agricultural strategies, discussed below, these problems can be rectified.

Crop rotation and diversification

Crop rotation and crop diversification work together to fight erosion, keep pests under control, and improve soil fertility. By planting different crops on the same field in successive years, it is possible to prevent the depletion of specific nutrients, which can occur if the same crop is cultivated in that field year after year. By including in the rotation cycle cover crops—such as legumes, which harbor symbiotic nitrogen-fixing bacteria—nitrogen and other soil nutrients can even be enhanced. Crop diversification has also been shown to be beneficial economically. It helps to stabilize and increase farm income while lessening dependence of the farmer on subsidies. Greater income stability will come through spreading the risk of price fluctuation over a number of different crops.

Soil conservation

Soil-protective methods such as reduced tillage, contour plowing, field terracing, and strip cropping all contribute significantly to a sustainable agricultural strategy. By minimizing the working of the soil before planting, removal of natural protective cover from the soil is prevented. This reduces erosion while maintaining crop yields, yet requires less investment of energy, labor, and equipment. Contour plowing, field terracing, and strip cropping all block the transport of soil by run-off. Together, these methods preserve the most essential agricultural resource, the soil, and also reduce farm expenses.

Natural strategies for pest control

Natural methods of pest control use diverse strategies to protect crops from pests—insects, weeds, pathogens, and other organisms. For instance, crop rotation and intercropping—the growing of two or more species in close proximity—are very effective in keeping insect populations under control. Another important strategy is to use natural predators to reduce insect populations.

The basic strategy of sustainable approaches to pest control is prevention. Healthy plants resist pests and pathogens. This approach extends the fundamental principles of Maharishi's Vedic Approach to Health to agriculturally important plants. By restoring and maintaining balance and vitality in crop plants, they are naturally less susceptible to insects and diseases. A similar strategy is used in weed control. Proper care and nurturing improves and enhances the health of the soil, the balance of nutrients, microorganisms and other soil components. Just as maintaining balance in the physiology through Maharishi's Vedic Approach to Health protects the human physiology from disease-causing pathogens, maintaining balance in the soil enables it to resist the damaging influence of weeds. Furthermore, just as Maharishi's Vedic Approach to Health takes into account physiological and seasonal cycles as part of its strategy to improve health, sustainable agricultural strategies take into account seasonal cycles and other timing considerations to minimize weed and pathogen damage and to maximize productivity. For instance, planting is often timed to give crop plants an advantage over weeds.

Research

Extensive objective research indicates that sustainable agricultural strategies make sense. For instance, a recent study found that corn and soybean yields actually increase after a transition to sustainable cropping systems. It further found that shifting toward sustainable farming methods through appropriate policy changes can raise agricultural productivity, reduce fiscal costs of maintaining farm incomes, and lower environmental costs in agriculture (49). Further research confirms the long-term profitability of sustainable agricultural practices (50).

Sustainable agriculture is clearly good for food production and farm economy, good for health and nutrition, and good for the environment. Sustainable methods have been available for years, yet they have not been widely adopted.

Why is this? It is due to choices that are made by farmers, by government leaders and, ultimately, by all of us. By enhancing our breadth of comprehension, creativity, and decision-making ability through Maharishi's Transcendental Meditation and TM-Sidhi program, we can play a pivotal role in implementing life-supporting agricultural practices.

Maharishi's Vedic approach solves agricultural problems

Applying Maharishi's Vedic approach in agriculture integrates natural sustainable agricultural methods that are free from chemical pesticides, herbicides, and other poisons, with the many technologies of Maharishi's Vedic Science. These include Maharishi's Transcendental Meditation and TM-Sidhi program, Yogic Flying, and Maharishi's Vedic Approach to Health. This approach works by dissolving stress, developing the consciousness of the farmer, and enlivening the collective consciousness of society.

These methods bring balance and integration to the farmer's life. In Chapter 3 we discussed how the Transcendental Meditation technique improves health (see Figures 7–12). We also presented scientific evidence that this technique improves comprehension (Figure 15), increases intelligence (Figure 16), enhances creativity (Figure 17), and improves decision-making ability. Through direct experience of the unified field of natural law in their own awareness, farmers develop the ability to make decisions that are spontaneously more in accord with the full range of the laws of nature governing agriculture. Thus farmers who adopt the Transcendental Meditation technique enhance their ability to organize and implement the complex network of factors influencing agricultural production.

The farmer as manager

The success of contemporary farmers depends not only on mastering those agricultural skills which are traditionally gained through "on the job training," but also depends on mastering a wide range of modern managerial and technical skills. Farmers must understand financial planning, market forces, and new and highly technical principles of soil chemistry and weed and pest management, as well as more traditional areas of knowledge, such as plant and animal breeding and the use and maintenance of agricultural equipment. Even in developing countries, the farmer is confronted today with new technologies and strategies for enhancing agricultural productivity.

Today's farmers must have the ability to assess short and long-term benefits, and possible harmful effects, of new agricultural

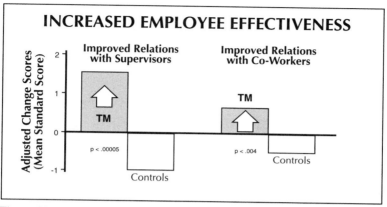

Figure 20. Increased Employee Effectiveness
Employees in the auto industry who learned Maharishi's Transcendental Meditation program showed significant improvements over a three-month period, in comparison to controls, in overall physical health, employee effectiveness, job satisfaction, and relationships at work and at home. They also showed reduced health problems, reduced anxiety, reduced insomnia, reduced fatigue, and reduced alcohol and cigarette consumption.

C.N. Alexander, G. Swanson, M.V. Rainforth, T.W. Carlisle, C.C. Todd, and R. Oates, "Effects of the Transcendental Meditation program on stress reduction, health, and employee development: A prospective study in two occupational settings," *Anxiety, Stress and Coping: An International Journal* 6: 245-262, 1993.

---

products. The decisions that they make are crucial, because these decisions influence not only the immediate productivity and profitability of their farms but also the health and welfare of those who consume their products, and the long-term viability of the soil for future generations. Farmers must understand the social and economic forces acting on agriculture, as well as the natural forces of sun, wind, rain, and soil. They must integrate all of these factors in deciding what to plant and when, and how best to care for their fields. If rains are late, or crops fail, farmers are challenged to find creative solutions.

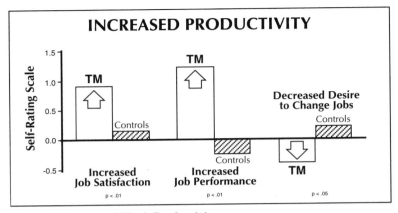

Figure 21. Increased Work Productivity
In this study, those practicing Maharishi's Transcendental Meditation program improved significantly in work productivity compared with members of a control group. Both job performance and job satisfaction increased, while the desire to change jobs decreased. People at every level of the organization, both executives and employees, benefited from practicing the Transcendental Meditation program.

D.R. Frew, "Transcendental Meditation and Productivity," *Academy of Management Journal* 17, 362–368, 1974.

Improving managerial effectiveness

The results described in Figures 7–17 of Chapter 3, suggest that the Transcendental Meditation technique would be a valuable tool for enhancing performance in the workplace. A number of studies have confirmed this hypothesis; the technique has been found to improve the productivity and effectiveness of both managers and employees. Although the research completed to date has not directly examined the use of this technique by farmers, use by professionals in a variety of other fields has been studied with similar results in all cases. This implies that it is valid to generalize the results of these studies to other professional groups, including farmers.

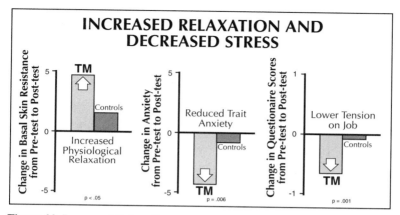

Figure 22. Increased Relaxation
Research was conducted in a Fortune 100 manufacturing corporation and in a smaller distribution-sales company showing that after three months managers and employees who regularly practiced the Transcendental Meditation program displayed more relaxed physiological functioning during a simple mental task than control subjects. In addition, over the three-month period the regular Transcendental Meditation practitioners showed a greater reduction in both state and trait anxiety and reduced tension on the job. Other measurements in this study revealed significant improvements on a wide range of physiological, psychological, and work-related measures in comparison to control subjects with similar positions in the same company.

C.N. Alexander, G. Swanson, M.V. Rainforth, T.W. Carlisle, C.C. Todd, and R. Oates, "Effects of the Transcendental Meditation program on stress reduction, health, and employee development: A prospective study in two occupational settings," *Anxiety, Stress and Coping: An International Journal* 6: 245-262, 1993.

Figure 20 presents a research study that examined the work performance of auto industry employees, showing that the Transcendental Meditation technique significantly enhances employee effectiveness, compared to controls. Figure 21 shows that work productivity is enhanced as well. This study found that several factors related to productivity, including job satisfaction and job performance, all improve in those practicing this technique.

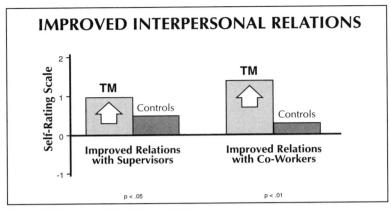

Figure 23. Improved Relations at Work
Relationships with co-workers and supervisors were significantly improved after an average of 11 months practicing the Transcendental Meditation program in comparison to control subjects. A study with several control groups (ref. 2) replicated these results and found that the degree of improvement was correlated with the length of time practicing Transcendental Meditation, and that administrators and employees both benefited from the Transcendental Meditation program.

Reference 1. D.R. Frew, "Transcendental Meditation and Productivity," *Academy of Management Journal* 17, 362–368, 1974.
Reference 2. K.E. Friend, "Effects of the Transcendental Meditation Program on Work Attitudes and Behavior," in *Scientific Research on the Transcendental Meditation Program: Collected Papers, Volume 1*, ed. D.W. Orme-Johnson and J.T. Farrow (Livingston Manor, N.Y.: MERU Press, 1977), 630–638.

The research presented in Figure 22 was conducted in the same Fortune 100 automobile manufacturing corporation studied in Figure 20. This research showed that the Transcendental Meditation technique increases physiological relaxation and reduces anxiety and tension on the job. These benefits were experienced by both managers and employees after just three months of regular practice of this technique. Since the ability to function without accumulating stress is a major determinant of long-term productivity and effectiveness, this result is further evidence that the

Transcendental Meditation technique can contribute to professional performance.

The ability to interact harmoniously and effectively with others is essential to the farmer, just as it is to other professionals. As illustrated in Figure 23, practice of the Transcendental Meditation technique significantly enhances the quality of interpersonal interactions. Both executives and employees experienced better relations at work as a result of practicing this technique, and the benefits increased over time. Whereas Transcendental Meditation practitioners reported that they felt less anxiety about promotion, their fellow employees saw them as moving ahead quickly. This indicates that people who practice this technique experience less anxiety and feel more confident about themselves.

Wise farming

As mentioned earlier, the professional performance of farmers is critical for the welfare of society and of future generations, as well as for their own financial success. Farmers manage one of our most essential resources—agricultural land—and they are responsible for producing our most critical commodity, the food we eat. The research reviewed above makes it clear that the Transcendental Meditation technique provides the farmer with valuable new resources—better health and vitality, greater creativity, and enhanced clarity of thinking. The result will be more appropriate and effective use of natural law, leading to wiser agriculture practices. Wise farming can produce more abundant and higher quality food, without resorting to technologies—such as agrochemical use and genetic engineering—that harm the environment and threaten human health.

A more intelligent and creative social climate

Group practice of the Transcendental Meditation and TM-Sidhi program provides additional benefits. Not only does this program enhance the life of the individual farmer, but it benefits society as well. This program dissolves stress in society and enlivens collective consciousness. As discussed in detail in Chapter 3, this leads to expanded comprehension and improved decision-making ability in society as a whole. It also creates a social climate in which leaders can make decisions that are more intelligent and creative. Figures 18 and 19 (Chapter 3) present a sampling of the 42 independent research studies that have repeatedly demonstrated the effectiveness of this approach over the last 20 years. This research indicates that this technology enables leaders and all others who influence agricultural policy and practice to discriminate more precisely among agricultural options and therefore choose safer and more effective approaches. In such a climate, sustainable agricultural practices will flourish, and dangerous high-tech approaches, such as genetic engineering, will be avoided.

This technology will also restore and maintain balance in society, removing natural and man-made conditions that hinder agricultural productivity, and ensuring support from those laws of nature that govern the environment, the soil, the seed and the weather, so that crops are abundant.

Summary

Agricultural genetic engineering is considered by some to be the great hope for feeding the world sometime in the future. However, this approach is dangerous to the environment and to human health. It offers only short-term benefits, and carries the risk of creating serious problems that will persist for generations. In contrast, sustainable agricultural methods, actualized through Maharishi's Vedic approach, can fulfill the same needs, and can do so immediately. This strategy can yield greater prosperity for our farmers and higher quality, healthier food for everyone. At the same time this approach will eliminate the widespread erosion, soil depletion, toxic pollution, and health problems that current agricultural methods bring and that agricultural genetic engineering will exacerbate.

Over the long term, this safer approach will enhance health and prosperity, and sustain agriculture that is more in harmony with the environment. The benefits offered by this approach are huge. In light of these benefits, there is no justification for subjecting society and future generations to the large risks of genetic engineering and other high-tech alternatives.

INVITATION

We invite all scientists who are concerned for the welfare of humanity to join us in research on consciousness, research into the unified field of natural law. This is a very fortunate time in the history of the development of science. Today, the study of this unified field of natural law is available, along with study of the specific laws of nature that administer the vast diversity of existence. The spontaneous use of the total potential of natural law is available through the subjective approach of Maharishi's Vedic Science, while the use of specific laws of nature is available through the objective approach of modern science.

The laws elucidated by modern science are localized expressions of the unified field of natural law. Gaining knowledge of this unified field—gaining knowledge of the whole—enables the individual to use knowledge of those isolated laws in a manner that is in harmony with all of nature. Use of specific localized laws can be spontaneously guided by knowledge of the whole. This is Maharishi's Vedic Engineering. Action that comprehends both the whole and the specifics can address any specific problem thoroughly, yet do so without creating unexpected side effects.

Thus, Maharishi's Vedic Science brings fulfillment to modern science and its technologies, finally making it possible to accomplish the aims that 300 years of purely objective scientific investigation has failed to achieve. This new unified approach will yield complete knowledge and technologies that can solve humanity's problems thoroughly and comprehensively. We no longer need to settle for knowledge and technologies that offer only partial solutions to a few problems and create more problems than they solve, while leaving many of humanity's most serious needs untouched. Come, and we will work together to create a world free from problems and fear, and free from the dangers that genetic engineering and other modern science-based technologies inevitably produce.

We also invite far-seeing leaders in government and industry to create a global research foundation that will support research on consciousness. Nurturing research on this most fundamental level of nature's intelligence will generate that knowledge which is essential for transforming the present, with its many, grave problems, into a healthy and secure future for all life on earth.

"Progress depends on new discoveries. The objective approach of modern science, having glimpsed the Unified Field of all the Laws of Nature, has invited scientists to transcend the objective approach and be guided by the theories of consciousness available in the Vedic Literature, and adopt a new experimental methodology using the subjective approach that is now readily available through my Transcendental Meditation, its advanced techniques, and the TM-Sidhi Program.

"Now, if progress in science is to continue, research has to be in the field of consciousness, the field of pure subjectivity, and the approach has to be subjective or self-referral.

"My Transcendental Meditation and TM-Sidhi Program offers that desirable subjective approach to research in the field of pure subjectivity, the self-referral Unified Field of Natural Law, whose scope ranges from point to infinity."

—Maharishi Mahesh Yogi, *Vedic Knowledge for Everyone*, (MVU Press, 1994)

REFERENCES

1. The genetic revolution. Elmer-Dewitt, P. *Time*, January 17, 1994.
2. Human gene therapy comes of age. Miller, A.D. *Nature*, 357, 1992.
3. Human gene therapy. Anderson, W.F. *Science*, 256, 808–813, 1992.
4. Human somatic gene therapy: progress and problems. Brenner, M.K. *Journal of Internal Medicine*, 237, 229–239, 1995.
5. Demand grows for 'positive' gene therapy. Verrall, M. *Nature*, 371, 193, 1994.
6. The human genome project: under an international ethical microscope. Knoppers, B.M., Chadwick, R. *Science*, 265, 2035–2036, 1994.
7. The ethics of gene therapy. Bolton, R.G. *Journal of the Royal Society of Medicine*, 87, 302–304, 1994.
8. Human gene therapy: Why draw a line? Anderson, W.F. *Journal of Medicine and Philosophy*, 14, 81–93, 1989.
9. Gene technique can shape future generations. Kolata, G. *The New York Times*, November 22, 1994.
10. Sequences and consequences of the human genome. Koshland, D. *Science,* 246, 189, 1989.
11. The human genome project: biological nature and social opportunities. Koshland, D. Paper presented at the Stanford Centennial Symposium, January 11, 1991.
12. Ethicists wary over new gene technique's consequences. Kolata, G. *The New York Times*, November 22, 1994.
13. Eosinophilia-myalgia syndrome and tryptophan production: a cautionary tale. Mayeno, A.N., Gleich, G.J. *Tibtech,* 12, 346–352, 1994.
14. Profit prescription—In marketing of drugs, Genentech tests limits of what is acceptable. King, Jr., R.T. *The Wall Street Journal*, January 10, 1995.
15. Drug alert!—What your doctor may not know: the undisclosed side effects of some prescription drugs could hurt or even kill you. *U.S. News and World Report*, January 9, 1995.
16. *Healthy People 2000: National Health Promotion and Disease Prevention Objectives.* U.S. Department of Health and Human Services, Public Health Service, 1990.
17. Actual causes of death in the United States. McGinnis, J., Foege, W. *J. Amer. Med. Assoc.*, 270, 2207–2212, 1993.
18. Allocation of resources to health. Stewart, C.T. *Journal of Human*

Resources, 6, 111, 1971.

19. Health habits, health care use and costs in a sample of retirees. Leigh, J.P., Fries, J.F. *Inquiry,* 29, 44–54, 1992.
20. *The Health Consequences of Smoking. A Report of the Surgeon General 1989.* U.S. Department of Health and Human Services Public Health Service, Office of Smoking and Health, 1989.
21. Impact of worksite health promotion on health care costs and utilization: evaluation of Johnson & Johnson's Live for Life program. Bly, J.L., Jones, R.C., Richardson, J.E. *Journal of the American Medical Association,* 256, 3235–3240, 1986.
22. *Scientific Research on the Transcendental Meditation and TM-Sidhi Program: Collected Papers Volumes 1–6.* MVU Press, Vlodrop, The Netherlands, 1976 to 1995.
23. In search of an optimal treatment for hypertension: A review and focus on Transcendental Meditation. Schneider, R., Alexander, C., Wallace, R. In Johnson, E., Gentry, W., Julius, S., Editors, *Personality, Elevated Blood Pressure, and Essential Hypertension.* Washington D.C., Hemisphere Publishing Corp., 1992, 291–312.
24. Medical care utilization and the Transcendental Meditation program, Orme-Johnson, D. *Psychosomatic Medicine,* 48, 493–507, 1987.
25. Reducing medical costs: the impact of Transcendental Meditation on government payments to physicians in Quebec. Herron, R.E., Hillis, S.L., Mandarino, J.V., Orme-Johnson, D.W., Walton, K.G. *American Journal of Health Promotion,* in press, 1995.
26. *Maharishi's Absolute Theory of Defense.* Maharishi Mahesh Yogi MVU Press, Vlodrop, The Netherlands, 1994.
27. *Human Physiology—Expression of Veda and the Vedic Literature.* Nader, T. MVU Press, Vlodrop, The Netherlands, 1994.
28. Consciousness as a field: The Transcendental Meditation and TM-Sidhi program and changes in social indicators. Cavanaugh, K.L., Glenn, T., Orme-Johnson, D.W., Mittlefehldt, V. *The Journal of Mind and Behavior,* 3, 67–104, 1988.
29. Test of a field model of consciousness and social change: The Transcendental Meditation and TM-Sidhi program and decreased urban crime. Dillbeck, M.C., Banus, C.B., Polanzi, C., Landrith III, G.S. *Journal of Mind and Behavior,* 9, 457–486, 1988.
30. The effects of the Maharishi Technology of the Unified Field: Reply to a methodological critique. Orme-Johnson, D.W., Alexander, C.N., and Davies, J.L. *Journal of Conflict Resolution,* 34, 756–768, 1990.
31. Test of a field theory of consciousness and social change: Time series analysis of participation in the TM-Sidhi program and reduction of violent death in the U.S. Dillbeck, M.C. *Social Indicators Research,* 22,

399-418, 1990.
32. The Clinton administration and the biotechnology framework. Union of Concerned Scientists, *Gene Exchange*, 4 (3/4), 6–11, 1994.
33. Ecological implications of using transgenic fishes in aquaculture. Hallerman, E.M., Kapuscinski, A.R. *ICES mar. Sci. Symp.*, 194, 56–66, 1992.
34. Herbicide-tolerant crops. Union of Concerned Scientists, *Gene Exchange*, 5(1) 6–8, 1994.
35. Environmental concerns with the development of herbicide-tolerant plants. Goldburg, R.J. *Weed Technology*, 6, 647–652, 1994.
36. OSU study finds genetic altering of bacterium upsets natural order. Hill, R.L. *The Oregonian*, August 8, 1994.
37. The effects of genetically engineered microorganisms on soil foodwebs. Holmes, M.T., Ingham, E.R. *Bulletin of the Ecological Society of America (Supplement)*, 75, 97, 1994.
38. Gene transplant speeds salmon growth rate. *The New York Times*, September 20, 1994.
39. Ecological risks of transgenic crops. Abbot, R.J. *Trends in Ecology and Evolution*, 9, 280–282, 1994.
40. Recombination between viral RNA and transgenic plant transcripts. Greene, A.E., Allison, R.F. *Science*, 263, 1423–1425, 1994.
41. Risk assessment: do we let history repeat itself? de Zoeten, G. *Phytopathology* 81, 585–586, 1991.
42. Plagiarism or protecting public health? Millstone, E., Brunner, E., and White, I. *Nature*, 371, 647–648, 1994.
43. Conflict of interest alleged in BGH approval. Puzo, D.P. *The Los Angeles Times*, April 21, 1994.
44. F.D.A. accused of improper ties in review of drug for milk cows. Schneider, K. *The New York Times*, January 12, 1990.
45. Sidelined by side effects. Green, D. *Financial Times*, February 23, 1995.
46. Data on release of genetically engineered organisms in the USA. Union of Concerned Scientists, *Gene Exchange* 5(2), 12, 1994; 5(1), 7, 1994.
47. Justus von Liebig *Agrikulturchemie*, 8. Auflage, 1865.
48. *Alternative Agriculture*. National Research Council, 1989.
49. *Paying the farm bill: U.S. agricultural policy and the transition to sustainable agriculture.* World Resources Institute, 1991.
50. Texas agriculture: growing a sustainable economy. Texas Department of Agriculture *American Journal of Alternative Agriculture*, 1, 1987; 153, 1990.

APPENDIX 1

MEDIA COVERAGE FROM NOVEMBER 17, 1994 NEWS CONFERENCE IN WASHINGTON, D.C.

At a press conference held in Washington, D.C., November 17, 1994, Dr. John Fagan announced his decision to return to the National Institutes of Health $613,882 in federal grant money and to withdraw grant proposals worth another $1.25 million. These grants would have supported research yielding information that could have been used for potentially dangerous genetic engineering applications. At that news conference he called for a 50-year moratorium on germ-line genetic engineering in humans and on the release into the environment of any genetically modified organism. He also urged his scientific colleagues to take safer, more effective research directions, and urged leaders to adopt the more life-supporting technologies of Maharishi's Vedic Engineering, instead of exposing society to the dangers of genetic engineering.

The news conference was attended by more than 30 members of the national and international press corps; the news filed by these media was featured in several hundred television and radio programs, and newspaper and magazine articles in the U.S. and worldwide.

Coverage included positive stories by *ABC* national television news, *CBS* national radio news, *The Washington Post* (reprinted in major newspapers throughout the U.S. and worldwide, including an editorial in the *International Herald Tribune*), *The Boston Globe, USA Today, Chicago Tribune, Associated Press* (reprinted in hundreds of newspapers nationwide), *Reuters* television and news agencies (reprinted widely around the world), *Worldwide Television News, ANSA* (Italian national news agency), Springer (German national news agency), South Korean and Taiwanese television and news agencies, *BBC Radio* and *Voice of America*.

Scientific and professional journals covering Dr. Fagan's announcement included *Science, Chemical and Engineering News, The Scientist, Chronicle of Higher Education, Chemistry and Industry*, and *Biotechnology Report*.

The Washington Post   Thursday, November 17, 1994

The Washington Post
AN INDEPENDENT NEWSPAPER

Genetic Engineering Breeds Costly Protest

by Rick Weiss

With all the complaining from scientists these days about the shortage of federal research money, John Fagan stands out as the last of the big spenders.

In a move that has become the talk of the scientific community, the 46-year-old molecular biologist is returning nearly $614,000 in grant money to the National Institutes of Health, while withdrawing his previously filed request for an additional $1.25 million in support. He is doing so to protest what he sees as rampant and unwise genetic tinkering with plants and animals and the release of these novel organisms into the environment.

"The benefits of genetic engineering have been oversold and the dangers have been underrepresented," said Fagan, who has received more than $2.5 million in NIH grant money since 1986.

Jerome Green, director of the NIH division of research grants and a 39-year veteran of the institutes, said this may be the first instance of a scientist returning grant money in protest, although a few have returned leftover moneys when their research was completed ahead of schedule. "In my memory I can't recall a similar situation," he said.

But Fagan is not your ordinary scientist. A Cornell University Ph.D. who spent seven years doing research in high-profile laboratories at the National Cancer Institute, he has been a professor of molecular biology at Maharishi International University in Fairfield, Iowa, since 1976.

The university was founded by Maharishi Mahesh Yogi, the guru of transcendental meditation. And although it is accredited to the Ph.D. level by the North Central Association of Schools and Colleges, its peculiar name and its emphasis on meditation and Indian traditional healing techniques (college classes begin and end each day with meditation sessions) has left scientists wondering what to make of Fagan and his views.

The most common question upon hearing the news, among those who don't know him: "Is this for real, or is he a flake?"

Fagan said the decision to return the money was not the result of some hokey spiritual epiphany (he is a practicing Christian who meditates regularly), but the culmination of a long evolution in his thinking that started in 1968. It was then, as an undergraduate at the University of Washington, that he learned transcendental meditation.

Immediately, he said, his grades improved, and he continued the practice throughout his graduate and postgraduate training, which focused on enzymes in the body that neutralize cancer-causing substances. By all accounts he was a model scientist.

After accepting a faculty position at MIU, however, he began to have qualms about the work that he and other "gene jockeys" were performing. He fretted about the kinds of changes that might be wrought on the human race by scientists' newfound ability to change the genetic blueprint.

Even more worrisome, he said, was the seemingly indiscriminate genetic manipulations being performed on plants and laboratory animals, and the potential for those new genetic combinations to spread to other creatures—a concern often expressed by anti-science gadflies like Jeremy Rifkin but rarely uttered by researchers themselves.

Fagan finally came to believe that even with the many layers of review by government and university biosafety committees, such research was irresponsible given how little is known about the long-term consequences of the release of these organisms into the environment.

"I'm concerned that we currently don't have enough data to predict the outcomes of these manipulations," he said, comparing scientists to 10-year-olds who think they are ready to drive a car. In a news conference scheduled in Washington today, he will call for a 50-year moratorium on the release of genetically engineered organisms. "I think he genuinely got worried about where the results were going to go," said Ruth Hubbard, a Harvard biology professor emeritus with whom Fagan has discussed his feelings. Scientists tend not to look far ahead, and when they do they look through rose-colored glasses, Hubbard said.

"We've all learned to write our grant applications in a way that gives the impression that it will benefit everyone," she said. "And I think Fagan just got to the point where he stopped believing that."

Others had a more predictable reaction. "I wonder if we can get some of that money," said one NCI researcher who used to work with Fagan but asked not to be identified.

In fact, NIH won't see a penny of the $613,882 that Fagan is returning, said Anne Thomas, NIH associate director for communications. Since the money was part of last year's budget, it must go back into the treasury's general fund—which is a little frustrating, Thomas said, for all the scientists who lost out in the frenzied competition for those dollars last year.

The Washington Post Monday, November 21, 1994

The Washington Post
AN INDEPENDENT NEWSPAPER

A Scientist's Qualms

Molecular Biologist John Fagan touched a nerve last week when he held a news conference to announce he was giving back $600,000 in federal research grants. His reasons—concern at the momentum of genetics research and belief that scientists should join him in backing a 50-year moratorium on certain types of commercial applications—tap into a lot of lay people's as yet unfocused worries about the possibilities biogenetic experimentation could unleash. Can a genetically engineered or mis-engineered organism "pollute" the environment? What about "Jurassic Park"?

These vague fears are based frequently on only partial understanding of the technology involved. And it is also true that the particular dangers Mr. Fagan stresses—so-called "germ-line" manipulation of the transmissible genetic material of organisms—are not necessarily so serious as to outweigh the usefulness of this work. Mr. Fagan's position, however, is not that of opposition to genetic manipulation per se but, rather, of caution in the face of the industry's momentum and, he says, an absence so far of coherent oversight. He suggests that biogenetic technologies are now "at the point where nuclear and chemical technologies were earlier in the century"—that is, on the optimistic brink of seemingly fabulous possibilities, but without any specific framework for avoiding the accompanying dangers.

What exactly might those dangers be? This is the point on which disputes among scientists are sharpest. It's also the central question for anyone trying to design a strategy to contain them. Mr. Fagan says he is most worried about environmental disruption, the "domino effect" that could hit ecosystems if a commercially engineered variation were released into the wrong environment, especially if its genes had been altered in a way that would make the change hereditary: "You can't recall a fish."

Other geneticists differ with Mr. Fagan's estimate of the level of danger inherent in some products that have been cleared by the patchwork regulatory structure. Bovine growth hormone occasioned a terrific fight but was eventually cleared by the Food and Drug Administration; so was the "FlavrSavr" genetically-enhanced tomato. Mr. Fagan's bigger point is that the excitement of these and other commercial possibilities could outstrip researchers' own caution and get beyond even the existing safety features. His gesture, and the attention it has drawn, could flash a useful yellow light on the stampede.

Science   Friday, November 25, 1994

FUNDING REVERSAL

Cancer Researcher Returns Grant

Ever since researchers spliced together genes from different organisms more than 20 years ago, researchers and ethicists have worried about where genetic engineering may lead, but few have gone as far as molecular biologist John Fagan. Last week, Fagan announced he is returning more than $600,000 in grant money to the National Cancer Institute (NCI) because he no longer wants to be a part of genetic research.

Fagan, chair of the chemistry department at Maharishi International University in Fairfield, Iowa, called for a 50-year moratorium on releasing genetically engineered organisms into the environment, pending further research, in a 17 November news conference in Washington, D.C. He also expressed concern about potential future manipulations of germ-line cells in both animals and humans. "There are people out there who think favorably on the idea [of]...potential eugenic applications," he said.

Fagan, age 46, has enjoyed 9 years of continuous funding from NCI; his latest grant renewal, for research into cancer susceptibility genes, came in September. He is also in the fourth year of a 5-year NCI Research Career Development Award. Although his former lab chiefs won't comment, two former colleagues call him a "competent" if not world-burning scientist.

Fagan is giving up his research to focus on what he considers to be a more fruitful activity: research on "traditional" medicine, specifically Indian Ayurvedic medicine, which he thinks holds more promise for disease prevention than does gene-splicing. A longtime practitioner of Transcendental meditation (TM)—the type advocated by Maharishi Mahesh Yogi, founder of this university—Fagan's career change has been germinating for some time. After getting his Ph.D. at Cornell University in 1977, he was a postdoc and later a senior staff fellow at the National Institutes of Health—where, says a former colleague, he also taught an informal course on TM. He moved to Iowa in 1984. Two years ago, he says, a "deluge" of media coverage convinced him that scientists had begun "to promote [genetic engineering] research in an unrealistic way."

The biomedical establishment is taking Fagan's defection in stride. NCI Director Samuel Broder issued a statement saying that his decision "could be in the best interests of all parties if he has lost enthusiasm for his own research." –Constance Holden

The Boston Globe — Wednesday, November 16, 1994

Biologist Returns US Grants to Protest Genetic Research

by Richard Saltus, Globe Staff

'This technology is going to put us in the same place that nuclear power did—we got burned not realizing the potential for side effects.'

John Fagan
Maharishi International University

A cancer researcher who brought a measure of scientific respectability to a small Iowa university, founded by the guru of the Transcendental Meditation movement, is returning nearly $614,000 in federal grants to protest the "grave dangers" of genetic research.

John Fagan, a molecular biologist at the Maharishi International University who once worked at the National Cancer Institute, said he is also withdrawing applications that could have brought another $1.2 million in grants to the university, which was founded by [the] Maharishi Mahesh Yogi.

His association with what many consider a fringe institution made Fagan's action less startling than it might have been. But it was still remarkable for a researcher who had succeeded in one of the most productive and intensely competitive scientific disciplines. He said he was hoping to call attention to doubts he has long harbored about his field.

"There's been a great overpromotion of gene therapy techniques," said Fagan in a telephone interview yesterday, "and this technology is going to put us in the same place that nuclear power did—we got burned not realizing the potential for side effects."

Fagan, 46, said he will focus on treating cancer and other diseases through a traditional Indian healing system known as Ayurveda. Already, he said, he has found promising "interim results" in experiments in which he gave a dietary substance, which he declined to identify, to rats with breast tumors. He said six out of eight cancers disappeared.

At a time when most scientists are competing desperately for scarce research funds, Fagan's action is unusual and possibly unique, said officials at the NIH, which awarded the grants.

"I know of people who've returned money because their work was completed sooner than they expected, but not for something like this," said Jerome Green, director of research grants.

The university is best known for its connection to the Maharishi, who brought transcendental meditation to Western popular culture through the Beatles in the 1960s and '70s. The campus, which has 650 students, has graduate programs in traditional fields like management, psychology and science, but also offers programs in Human Consciousness and Creative Intelligence.

The school, founded in 1971, had little presence in molecular biology until Fagan arrived in 1984 after 7 years at the National Cancer Institute. He used gene-splicing and other genetic tools in probing the cytochrome p450 system, a group of enzymes in the liver that help destroy toxic drugs and harmful byproducts of metabolism.

Within two years, Fagan, who has a Ph.D. from Cornell, landed the first of NIH grants totaling $2.5 million for the 27-member research group he built—a great coup for the Iowa university.

"It was a great breakthrough," said Robert Keith Wallace, executive vice president, in an interview. "We had to have a lot of visits and a lot of reviews—it wasn't easy for them to trust us as a research university. We have other professors who've gotten federal grants, but he's been the key."

Fagan said he had long had concerns about the potential misuse of genetic technology, which led him to "take stock" of his work over the past year or two.

"My work was heading toward the discovery of genes that cause people to become susceptible to cancer," he said. "But that knowledge can't be used to cure someone. The only way it could be used would be in germ-line gene therapy"—meaning replacing mutant genes with healthy ones that would be inherited by the individual's children "to protect the next generation."

Such measures, said Fagan, raise "the problems of eugenics and the risk of side effects and new genetic diseases." He also warned that genetic technology could be used by "terrorists or petty tyrants" to create weapons for "biotech terrorism."

Nelson Wivel, who heads NIH's recombinant DNA Advisory Committee that regulates gene therapy, said the committee does not consider proposals for germ-line gene therapy experiments. While he did not rule out reexamining that policy in the future, he said the technical hurdles are so formidable "that it would be a long, long time before anyone would even consider" adding genes to human reproductive cells.

Chicago Tribune   Friday, November 18, 1994

Scientist Returning Grant, Opposes Genetic Engineering

Associated Press

WASHINGTON—A scientist with the Maharishi International University is taking a stand against genetic engineering studies by withdrawing from research projects and returning more than $600,000 in federal grants.

John Fagan, a professor of molecular biology at the institution in Fairfield, Iowa, announced at a news conference Thursday that he was returning to the National Institutes of Health $613,882 in cash grants and will withdraw from grant proposals that were expected to be worth $1.25 million.

Fagan has been conducting genetic research, but has become convinced that such studies could eventually be harmful to the welfare of humanity. In a statement, he said that by returning the money he hopes to sound a warning about what he sees as serious dangers of genetic engineering.

Scientists are manipulating the genes of plants and laboratory animals. Some manipulated plants, such as a tomato that was genetically altered to control ripening, have been approved for general use.

Some human cell genes have been altered for treating specific disorders, but regulations formulated by NIH specifically forbid manipulations of genes that could affect inherited characteristics.

"I'm concerned that we currently don't have enough data to predict the outcomes of these manipulations," Fagan told *The Washington Post*. He said there should be a 50-year moratorium on developing genetically engineered plants and animals.

Anne Thomas, a spokeswoman for the NIH, told the *Post* that any money returned by Fagan will go back into the general federal treasury, since it was in last year's budget, and will not be available for use this year by the NIH.

USA Today   Fri./Sat./Sun., Nov.18-20, 1994

Gene scientist, with warning, spurns grants

From staff and wire reports

A scientist involved in DNA research announced Thursday he is returning a $613,882 grant to the National Institutes of Health because genetic manipulation may have "dangerous consequences" he says scientists today can't even begin to predict.

"To make such changes based on highly incomplete data is to make irreversible decisions that will affect all generations to come," says John Fagan, a professor of molecular biology at Maharishi International University in Fairfield, Iowa. He says he also is withdrawing proposals for other DNA research grants worth more than $1.25 million.

He called for a 50-year ban on releasing genetically engineered organisms into the atmosphere, "because we don't have enough data to predict the outcomes...Until we do, we should not allow genetic technologies to be applied."

Fagan, 46, holds a doctorate from Cornell University and worked for seven years as a researcher at the National Cancer Institute. He says he plans to pursue "safer, more productive" health research, including approaches such as Ayurveda, a traditional system of preventive medicine from India, popularized in the West by Maharishi Mahesh Yogi.

Hindustan Times   29 November, 1994

Researcher returns grant

Citing the grave dangers of genetic engineering, Dr. John Fagan, a recognized DNA researcher, will give back to the National Institutes of Health, $613,882 in cash and withdraw other grant proposals for high priority research projects worth more than $1.25 million that would have been used for DNA research. The results of the research, he said, could have contributed to harmful genetic engineering applications.

Dr. Fagan, 46, a Cornell University-trained molecular biologist and Professor of molecular biology at Maharishi International University in Fairfield, Iowa, held a news conference in Washington to alert the public to the "dangerous consequences of genetic engineering" and to call for a 50-year moratorium on germ-line genetic manipulations. (In germ-line therapy, new genes are introduced into the DNA of sperm, eggs, or very early embryos. Changes introduced into the germ cells are passed on to all future generations. Thus accidents and unpredicted side effects will be perpetuated indefinitely.)

Dr. Fagan has conducted DNA research for 23 years. He currently heads a 27-member research group at MIU's Molecular Biology Laboratory. Since 1986 the lab has received federal grants totaling more than $2.5 million from the NIH.

"The risks of genetic engineering are serious and growing," Dr. Fagan said. "The trend is now toward increasingly rapid application of gene therapies and genetic engineering methods without proper consideration of their impact." He said the long-term and cumulative effects of germ-line genetic engineering in any species will lead to irreversible and highly unpredictable effects on the quality of life for the whole planet.

"The different organisms that populate earth do not live in isolation. Their lives are intertwined in countless ways. Any change made in any organism will, over the long term, influence this whole fabric of life.

"In this extended time frame, it is not possible for any researcher to predict the consequences of a change in the genetic makeup of any species. To make such changes based on highly incomplete data is to make irreversible decisions that will affect generations. We don't have sufficient information to make these decisions responsibly. Until we do, we should not allow genetic technologies to be applied," he said.

Five European countries have already banned germ-line genetic engineering in humans.

Dr. Fagan now plans to pursue "safer, more productive directions of research" that offer greater potential for improving health and fighting disease, including research on the Vedic approach to health, India's traditional system of natural preventive medicine as brought to light by Maharishi Mahesh Yogi.

The China Post — Saturday, November 19, 1994

THE CHINA POST
COMMENTARY
Saturday, November 19, 1994

Scientist gives back research funds

Reuter

Washington, Nov. 17— Warning that science must not tamper with nature, a U.S. molecular biologist said on Thursday he was giving back a US$600,000 government cancer research grant and refusing US$1.25 million more.

John Fagan told a news conference he was getting off the high-tech, genetic engineering treadmill and going back to ancient Asian wisdom for curing 20th century diseases.

"I'm not talking about whacked-out stuff here," said the 46-year-old researcher, like the Beatles before him a believer in Maharishi Mahesh Yogi's transcendental meditation.

But it was not the Maharishi's meditative answers to spiritual life that changed his mind about modern science.

He said he had thought it over for two years and decided that there is something seriously wrong with tinkering too much with the genetic makeup of plants and animals and releasing organisms into the environment.

Fagan announced he is returning a US$614,000 grant to the U.S. National Institutes of Health, awarded him in September, and withdrawing a request for an additional US$1.25 million.

No one at NIH, which oversees millions in grants for promising research each year, could recall anyone ever returning grant money as a protest.

Fagan said his future is in the holistic approach of traditional Asian medicine. He intends to devote his research efforts to pursing the medical wisdom of the Orient rather than continuing as a "gene jockey."

He called for his fellow scientists to impose a 50-year ban on releasing genetically engineered organisms into the atmosphere "because we don't have enough data to predict the outcomes of these manipulations."

The Cornell University Ph.D. and former researcher at the National Cancer institute is now a professor at Maharishi International University in Fairfield, Iowa.

"We should move more prudently on genetic engineering," Fagan said.

Chemical & Engineering News December 19, 1994

GOVERNMENT
Researcher returns NIH funding in protest

For the first time, a scientist is returning grant money to the National Institutes of Health to protest what he sees as dangers associated with his field of research.

John B. Fagan, chairman of the department of chemistry and professor of molecular biology at Maharishi International University, Fairfield, Iowa, is giving back almost $614,000 to the National Cancer Institute (NCI) and withdrawing his request for an additional $1.25 million in support to protest what he sees as unwise genetic manipulation of plants and animals.

Fagan, who leads a 27-member research group that works on cancer susceptibility genes, is calling for a 50-year moratorium on the release of genetically engineered organisms into the environment. Since 1986, he has had continuous funding from NCI, receiving grants worth more than $2.5 million.

Fagan is not opposed to biotechnology research itself. But "The applications of our research create more problems than they solve, and those problems are serious," he says. "I also believe that our research generally promotes the troubling trend toward expanded use of fundamentally dangerous genetic technologies in medicine and other areas," he adds.

"It would be ethically inconsistent for me to continue a research program that leads to the use of technologies that I consider a threat to the health and well-being of society."

Since biotechnology research began 20 years ago, some scientists have warned about its potential dangers and applications, but no one has gone so far as to return grant money. For example, when Jonathan R. Beckwith, a molecular biologist at Harvard University, first isolated a human gene in 1969, he gave a speech saying he was concerned that this might lead to manipulation of human germ-line cells—which produce heritable traits.

Today, many ecologists and some biologists are still concerned that bioengineered organisms could upset ecological systems or that manipulation of human germ-line genes could have unforeseen negative heritable consequences. However, in their public statements at least, molecular biologists who do research in this area say the field is adequately regulated and has great potential for improving food production and for preventing and curing disease. Federal regulators contacted by C&EN generally believe that current rules and guidelines for biotechnology are strict enough to prevent potential problems.

The Des Moines Register   Sunday, December 11, 1994

Des Moines Sunday Register

REGISTER EDITORIALS

When science gets ahead of ethics

Like A-bomb research, genetic experiments raise moral questions.

Microbiologist John Fagan of Maharishi International University in Fairfield waved the "caution" flag over the scientific speedway last month, hoping to slow the dangerous race towards the creation of artificially crafted human beings (See the article on Page 2C of this section).

Genetic engineering—the hottest topic on the worldwide science agenda—is following ill-defined goals toward unknown consequences and without proper research safeguards, says Fagan. Manipulators working with the tiniest bits of organic matter are literally scripting the future, without proper knowledge of or concern for the outcome, Fagan says. He is halting his own genetic-engineering research, rejecting a federal grant and withdrawing applications for more—thereby fanning the coals of a simmering debate.

Fagan has chosen to confront an issue too frequently avoided by those most directly involved. An imaginary wall of separation between science and ethics has enabled scientists to beg off when questioned as to ethical implications of their discoveries. "We just do the research," the rationale goes, "we don't control the results."

Probably the most obvious clash of science and ethics came with development and use of the atomic bomb. Some of those involved are said to have viewed their work strictly as the pursuit of physics, divorced from the messy business of mass death. Others endorsed the pursuit both for its contribution to physics and to the war effort. Dropping the bomb would shorten the war, cost a hellish number of lives but save even more, and seal victory for the side of righteousness.

But in the years immediately following the end of World War II, most of America's top atomic scientists argued against developing the far more deadly hydrogen bomb. The exception was Edward Teller, who had the backing of the generals and the ear of influential politicians. He promoted and proudly claimed parentage of the most destructive device ever known.

If the history of nuclear-weapons development reveals an ethical ambivalence on the part of many scientists involved, there were some who were unswerving in their convictions, and who just may have saved civilization. That their names are little known is not surprising; they were part of Hitler's Germany.

The major hero of the story is Werner Heisenberg, Germany's top atomic scientist. It was fear of the possibility that he might hand Hitler an atomic bomb that launched the Manhattan Project in America. But Heisenberg and other top German physicists, it turned out, were quietly dragging their feet, telling Nazi leaders that building an A-bomb would take forever, was too large a task and might not work. In the words of one conspirator, "No one of us wanted to lay such a weapon in the hands of Hitler."

"Thus they betrayed their government, but served humanity well," writes Jeremy Stone, president of the Federation of American Scientists. Had Heisenberg and his fellow atomic researchers simply pursued science for science's sake and left the use of their discoveries up to the politicians, Hitler might have showered them with medals, then showered London with nuclear-tipped V-2 rockets.

In time, human genetic engineering—like the near worldwide spread of atomic weaponry—will become reality. What Iowa's John Fagan is asking of his fellow microbiologists is that they agree to slow down and let ethics catch up. Set the ethical rules for genetic engineering before turning every lab loose to mess with the blueprints of coming generations. Wait for the research that shows the geneticists how to avoid monstrous mutations.

Such rules can't bind all those who have the ability to do genetic manipulation. But they should make for more orderly progress with less chance of accidental consequences.

APPENDIX I

The Des Moines Register Sunday, December 11, 1994

Des Moines Sunday Register

OPINION

Genetic engineering's sinister side
Why an Iowa scientist returned a big grant

by Bill Leonard

There's a sinister side to the exciting and romantic new field of genetic engineering that the public is little aware of, says an Iowa scientist. And he no longer wants to be a part of it.

As a result, while researchers across the country search for ways to pry scarce dollars out of the federal government, professor John Fagan of Maharishi International University in Fairfield is returning his latest grant—$614,000—and withdrawing applications for another $1.25 million. A spokesman for the National Institutes of Health, which controls the search money, said Fagan's action could be a first.

Fagan, a molecular biologist, told a Washington news conference he is "taking an ethical stand." Germ-line genetic engineering—that which alters the reproductive or "germ" cells—can alter the nature of future human generations in ways science cannot predict. Side effects could show up that would be passed down to endless generations.

Such as?

Physical defects, weakened immune system, you name it.

"Who has the right to manipulate the genes of someone not even born yet?" he asks.

Manipulate them how?

"You want a Mozart? You want a football player?" It could be attempted, through insertion of the proper genes into the DNA—the chemical carrying genetic properties—of the embryo.

"You want a 7-foot tall son? That one we could do tomorrow," Fagan said.

"All you have to do is give him an extra copy of the growth hormone."

But in the process you could cause a mutation that made him particularly susceptible to cancer.

"He could be 7 feet tall, but he might have a weak heart, or weak joints. Or a weak mind."

So could his children, and theirs. The weaknesses could dog his offspring through endless generations.

"We don't know what else we could be doing," Fagan said, "and that is irresponsible."

He likened it to blasting the side of a barn with a shotgun. You know you'll hit the barn, but you don't know what else.

Genetic engineering, in short, becomes genetic pollution, the biologist said. And it not only can cause problems, it already has.

Fagan cites the example of coho salmon, engineered to grow bigger. It works. Some salmon are from three to 15 times their normal size. They also show genetic defects. And not all the implications are understood.

Salmon that used to be the prey of larger fish are now their predators. The intricate and gradually developed balance of nature is abruptly changed, too fast for nature to adjust. The sudden introduction of a new (man-made) species can have vast and important impacts all up and down the food chain.

Genetic engineers work to produce crops resistant to herbicides, so the chemicals will wipe out only the "bad" plants. But the accidental cross-pollination with wild species means we'll also have herbicide-resistant weeds—and in time, the toughies will take over, meaning most weeds will develop resistance. Farmers will then face greater problems than they started with.

While science has learned the art of genetic manipulation, it's far from foolproof, Fagan notes. You could tamper with the wrong gene, causing changes you neither wanted nor contemplated.

Most animal genetic engineering research involves mice. Anyone receiving federal money to pursue the research pledges not to manipulate the genes of humans. But not all research is federally financed.

"I doubt that anyone capable of doing it, would do it," Fagan said. But already, more than 1,000 scientists have the capability. And that field will expand.

Fagan who earned his doctorate at Cornell University in upstate New York, began work in genetic engineering in 1977 and joined the MIU faculty in 1984. His philosophical turnaround did not result from a road-to-Damascus conversion; his concerns for the dark side of the search developed over time. As long ago as 1983, he lectured a United Nations conference in Seoul, South Korea, on the ethics of genetic tampering. (Fagan's biographical material mentions addresses to scientific symposiums nationwide and in nine foreign countries.)

Fagan's work has centered on identifying genes associated with cancer susceptibility. When such genes are located among the 100,000 or so in each human cell, people carrying the faulty genes could be warned to avoid the usual risks associated with the disease.

"On the other hand," he said, "that information in the hands of an insurance company or an employer could be used against the person rather than for them."

"If I were to take 2 percent of the money now spent on genetic engineering and use it for research into preventive approaches in all areas of health, we'd be far ahead." The preventive approaches would involve such mundane factors as diet, exercise and drugs.

Meanwhile, "I'm calling for a 50-

year moratorium on germ-line research in humans and animals. That doesn't mean stopping research. What we need is more research so we can predict what's going to happen."

"In its beginning, genetic research was guided by some very elegant scientific methodologies," Fagan said.

"But now," he said, "at the point of implementation, science is being tossed out the window. What's guiding implementation are economic and some political pressures."

Fagan's decision cost MIU a healthy hunk of research money. How did MIU President Bevan Morris react?

"He swallowed hard," the biologist said. "But he recognized I was justified in doing this; that it was the right thing to do. And because of that, MIU has stood behind me."

In a statement supporting Fagan's decision to reject federal financing, Morris praised his "brilliant research...as reflected by the many research grants he has received," but noted that it is the responsibility of MIU and all universities "to do research that is only supportive of life and health, now and for future generations."

Fagan and MIU are far from alone. Thousands of scientists are involved in gene research, and many of those Fagan talks to share his concerns. Among them:

Martha Crouch, associate professor, Indiana University: "If only more experts would use their privileged knowledge to 'just say no' to dangerous technologies, as John Fagan has done, we would be on the way to a saner, safer world."

Philip Regal, a University of Minnesota professor: "It is a pity that the public has not been accurately informed about these critical issues."

Stuart Newman, professor, New York Medical School at Valhalla: "It is to Dr. Fagan's great credit that he has foreseen the social and medical harm...and decided to turn his efforts to educating the public."

Given the frightening consequences of misuse of the new technology, it's important that the public get up to speed on the issue in a hurry.

MAHARISHI INTERNATIONAL UNIVERSITY

NEWS RELEASE

DNA Researcher Takes Ethical Stand Against Genetic Engineering— to Give Back $613,882 NIH Grant

and Withdraw Other Grant Proposals Worth $1.25 Million

Cites dire consequences of new DNA technologies; calls for a 50-year moratorium on germ-line genetic manipulations

NEWS CONFERENCE
Thursday, November 17, 10:00 a.m.
Capital Hilton Hotel, Senate Room, 16th & K Streets, NW, Washington, D.C.

Citing the grave dangers of genetic engineering, Dr. John Fagan, an internationally recognized DNA researcher, will give back to the National Institutes of Health $613,882 in cash and withdraw other grant proposals for high priority research projects worth more than $1.25 million that would have been used for DNA research. The results of the research, he said, could have contributed to harmful genetic engineering applications.

Dr. Fagan, 46, a Cornell University-trained molecular biologist who is Professor of Molecular Biology at Maharishi International University in Fairfield, Iowa, will announce his decision during a news conference on Thursday, November 17, 10:00 a.m., at the Capital Hilton Hotel, Senate Room, 16th & K Streets, NW, Washington, D.C.

(MIU is accredited to the Ph.D. level by the North Central Association of Colleges and Schools).

Dr. Fagan said that he is holding the news conference to alert the public to the "dangerous consequences of genetic engineering" and to call for a 50-year moratorium on germ-line genetic manipulations. (In germ-line therapy, new genes are introduced into the DNA of sperm, eggs, or very early embryos. Changes introduced into germ cells are passed on to all future generations. Thus accidents and unpredicted side effects will be perpetuated indefinitely.)

Dr. Fagan also plans to mobilize support among scientists to pursue new directions of research that are safer and more productive.

Dr. Fagan has conducted DNA research for 23 years. He currently heads up a 27-member research group at MIU's Molecular Biology Laboratory. Since 1986 the Lab has received federal grants totaling more than $2.5 million from the NIH.

Dr. Fagan conducted his post-doctoral research at the NIH, and is recipient of a prestigious Research Career Development Award from the National Cancer Institute. He is a frequent speaker at international scientific conferences, has served on NIH peer-review committees for research grants, and is an editorial advisor and reviewer for scientific journals.

Dr. Fagan's grants from the NIH support research on two genes that are blueprints for enzymes, called cytochromes P450, which are involved in carcinogen and toxin metabolism. He said that this research itself is safe, but basic information and techniques emerging from his work could contribute to harmful genetic engineering applications.

Dangers of Applying Genetic Engineering without Proper Consideration

"The risks of genetic engineering are serious and growing," Dr. Fagan said. "The trend is now toward increasingly rapid application of gene therapies and genetic engineering methods without proper consideration of their impact."

He said that the long-term and cumulative effects of germ-line genetic engineering in any species will lead to irreversible and highly unpredictable effects on the quality of life for the whole planet.

"The different organisms that populate earth do not live in isolation," Dr. Fagan said. "Their lives are intertwined in countless ways. All of these species have co-evolved over a huge span of time, and this process is continuing. Any change that we make now in any organism will, over the long term, influence this whole fabric of life.

"In this extended time frame, it is not possible for any researcher to predict the consequences of a change—no matter how minor it seems at the present—in the genetic makeup of any species. To make such changes based on highly incomplete data is to make irreversible decisions that will affect all generations to come. We don't have sufficient information to make these decisions responsibly. Until we do, we should not allow genetic technologies to be applied," he said.

Five European countries have already banned germ-line genetic engineering in humans.

Heightened Risk of Biotech Terrorism

Dr. Fagan also cited his concerns about the heightened risk of "biotech terrorism."

"All the tools required to genetically engineer a biological weapon are freely available and unmonitored on the open market here in the U.S. and abroad," Dr. Fagan said. "For a tiny fraction of the cost needed to develop a nuclear weapons program, any terrorist or petty tyrant could use recombinant DNA techniques to develop an effective biological weapon, such as a pathogenic organism or virus, that could threaten the whole of humanity. Biotech terrorism is a catastrophe waiting to happen."

Recombinant DNA Methods Fail to Deliver Health Promise

Since the late 1970s Dr. Fagan has been using recombinant DNA methods, pursuing the promise of bringing practical benefits to medicine. He recently "took stock" of the results—not only his own findings but also the results produced by the whole of the biomedical research community and, he said, he is not convinced the investment has paid off.

"For example, recombinant DNA methods have led to significant progress in understanding fundamental mechanisms of carcinogenesis. But they have been far less effective in generating practical procedures for preventing and treating cancer.

"This lack of substantive progress, combined with my growing concerns about the dangers of genetic manipulations, have led me to this decision, where I must, in good conscience, return the NIH grants that were awarded for my DNA research," Dr. Fagan said. "I urge other researchers in the field to do the same."

Safer, More Productive Directions of Research

Dr. Fagan now plans to pursue "safer, more productive directions of research" that offer greater potential for improving health and fighting disease, including research on India's traditional system of natural preventive medicine, the Vedic approach to health. A large body of scientific data has already documented the effectiveness of the Vedic approach, as reformulated by Maharishi Mahesh Yogi, for preventing and treating a broad range of disorders "that genetic engineers only aspire to address." These include heart disease, hypertension, cancer, aging, arthritis, substance abuse, violent behavior, anxiety, and depression.

He said that a vast difference exists between the genetic engineering approach to health and the Vedic approach to health.

"Applying genetic engineering carries the risk of dangerous side effects because that body of knowledge is incomplete—we know only fragments of the story. On the other hand, the Vedic approach produces no harmful side effects because it is inherently holistic. It deals with the integrating intelligence that underlies all matter, whereas genetic engineering deals with the isolated point values of matter itself. While billions of dollars are being spent in developing high-tech, high-risk approaches that, at best, have very restricted applicability, research shows that the Vedic approach is highly effective, preventing and treating disease naturally and cost-effectively.

"I look forward to this new direction of research," Dr. Fagan said. "It will make a safer, more powerful contribution to improving health and fighting disease."

MIU President Supports Dr. Fagan's Stand

Dr. Bevan Morris, President of Maharishi International University, said that he supported Dr. Fagan's decision to return grant funds to the NIH.

"Dr. Fagan is a brilliant scientist who has made a principled decision based upon real concerns about the dangers of recombinant DNA research," Dr. Morris said. "I support him in his stand.

"I am extremely excited that Dr. Fagan has decided to research the field that underlies all matter—the field of consciousness, or intelligence. His work will further the discovery of Tony Nader, M.D., Ph.D., who found the total potential of natural law, as embodied in Veda and Vedic literature, to exist in human physiology.

"Dr. Fagan's new research promises profound applications for preventing and treating disease and promoting health and enlightenment for the individual and society," Dr. Morris said.

APPENDIX I

FACT SHEET FOR REPORTERS

Recombinant DNA Research Program

Background and explanation of why I am terminating this program and returning grant money to the NIH

OVERVIEW OF RESEARCH

Our research program has focused on two genes that are important in protecting the body from a host of toxins that are imbibed with our food, water and air. These genes are the blueprints for enzymes, called cytochromes P4501A1 and P4501A2, which metabolize, or break down, environmental toxins. This area is particularly relevant to cancer, since many of these toxins are carcinogens. In some cases these P450 enzymes break down, or detoxify, these compounds thus protecting the body from their harmful effects. In other cases, the P450s do just the opposite—they chemically alter, or activate, the carcinogens thus making them even more potent, more carcinogenic. The direction that is taken, toward detoxification or toward activation, seems to be partially genetically determined. That is, some people seem to be genetically predisposed to detoxification; others to activation. This may partially explain why some people are more susceptible to cancer than others. Our research has focused on identifying genes responsible for individual differences in sensitivity to chemical carcinogens—cancer susceptibility genes. (A more technical discussion of the rationale behind our work and our recent research results can be found on the third page of this Fact Sheet.)

POSSIBLE APPLICATIONS

Once a cancer susceptibility gene has been identified, how could that knowledge be applied? There are three possible applications:

1. A gene technology-based diagnostic test could be developed for identifying individuals who are abnormally susceptible to chemical carcinogens or to environmental toxins.

2. Gene replacement therapies could be developed to correct the genetic basis for that increased susceptibility.

3. Genetic engineers could use for other purposes the basic information on gene regulatory mechanisms and refinements of research methods that would emerge from this research.

CONCERNS ABOUT HARMFUL OUTCOMES

In this brief statement, it is not possible to do justice to the complexities and subtleties of these issues. Here I will consider only the most obvious effects of the possible applications of our work.

1. Dangers of diagnostic testing: Misuse outweighs payoff. Currently, there is no procedure available that would correct a genetic defect identified by this diagnostic test. Thus, this analysis would provide little or no useful information to the person tested, except as a warning to avoid exposure to potentially harmful substances. On the other hand, the harmful outcomes from this genetic test would be much more widespread. For example, an insurance company or a potential employer could use this information to avoid insuring or hiring a person who might get cancer. Such an application of our research would adversely impact the lives of many people and, therefore, I believe that this misuse far outweighs any potential payoff.

2. & 3. Dangers of gene replacement: Creating new genetic diseases. Replacing or correcting a gene responsible for increased cancer risk would, in principle, reduce susceptibility to cancer. While this is not technically feasible at present, it will be in the future. However, to protect against cancer such a procedure would have to correct the genetic information in each of the countless cells of the body or, at least, in every cell of those tissues that are sensitive to a particular carcinogen. This is impractical by somatic cell genetic manipulations. Therefore, the only way that the results of our research could be applied to prevent cancer would be through hazardous germ-line gene therapy—a procedure that has already been banned for use in humans by five European nations. If successful, this would reduce by 15 to 20% the likelihood that a recipient would get cancer. However, the potential negative effects are far greater and long lasting, such as accidental genetic defects or other unanticipated negative side effects. The most serious danger of germ-line therapy is that errors and side effects would not only threaten the individual who undergoes this therapy, but also all of his or her progeny for all subsequent generations. In essence, this procedure carries the real risk of creating a new genetic disease because every genetic manipulation is a potentially mutagenic event. In addition, the use of germ-line methods raises the specter of eugenics. Clearly, the potential benefits from this therapy—a small partial reduction in susceptibility to carcinogens—do not justify these serious, long-range risks. This is especially true when other currently available preventive means offer far greater protection.

Decision to terminate research program and return NIH grants

It is clear that the applications of our

research create more problems than they solve, and those problems are serious. I also believe that our research generally promotes the troubling trend toward expanded use of fundamentally dangerous genetic technologies in medicine and in other areas. It would be ethically inconsistent for me to continue a research program that leads to the use of technologies that I consider to be a threat to the health and well-being of society. Therefore, I have decided to terminate this research program and return to the NIH the grants awarded to us to pursue this research.

THREE PURPOSES FOR MY ACTION

To expect that my withdrawal from this research would, in itself, significantly slow progress in genetic engineering, would be highly unrealistic. If current trends continue, others will inevitably carry out the research that we have chosen to stop. However, I believe this action can generate momentum in a new direction. My primary purposes are:

1. Ethical stand. I am taking an ethical stand. I do not wish to contribute to something that I feel is harmful to the welfare of humanity.

2. Public warning. I want to sound a warning against the grave dangers of genetic engineering.

3. New directions of research. I want to let serious scientists know that there are safer and more productive research directions that are open to them.

TECHNICAL SUMMARY OF RESEARCH RATIONALE AND FINDINGS

Carcinogen metabolism and the balance between activation and detoxification are influenced strongly by the amounts or activities of P450s in the cell. Cellular levels of P4501A1 and P4501A2 (the P450s that we study) are, in turn, determined primarily by the rate of the first step of P450 gene expression, transcription. Thus, understanding how the expression of these genes is controlled in the cell may provide clues to the identity of the gene or genes responsible for individual differences in sensitivity to chemical carcinogens—cancer susceptibility genes. These could be the P450 genes themselves, or they could be genes encoding other proteins that function in the cell to control the expression of the P450 genes, such as nuclear factors that control the transcription of the P450 genes.

Based on this line of reasoning, our research has focused on gaining in-depth, basic knowledge concerning the regulation of the expression of the P450 genes. Evidence from our laboratory and other laboratories indicate that the expression of the genes encoding P4501A1 and P4501A2 is induced by a variety of polycyclic aromatic carcinogens and that both transcriptional and post-transcriptional mechanisms are involved, although transcriptional mechanisms are of primary importance. Therefore, our research focused primarily on this topic, examining both the cis-acting sequences and trans-acting factors that mediate the transcriptional response of the P4501A1 and P4501A2 genes to polycyclic aromatic carcinogens and to other, possibly liver-specific, effectors. We also studied the influence of chromatin structure on the transcriptional regulation of these genes. The gene regulatory mechanisms elucidated by this work could provide clues pointing to genes that may influence cancer susceptibility. For instance, our work recently brought to light a new family of proteins that play an important role in regulating P4501A1 gene transcription. The next step in our work would have been to follow up on this clue, and look for correlations between cancer susceptibility and specific polymorphisms in the genes encoding these regulatory proteins. Such correlations would, in turn, suggest that a particular gene may be a determinant of carcinogen sensitivity, a cancer susceptibility gene.

QUESTIONS FROM THE PRESS

Press Conference, Washington, D.C., November 17, 1994

J=John Fagan; R=Reporter

R: I'd like to know what your last grant was and when you got it. It seems like you changed your mind awfully fast. Also, do you plan to continue doing research?

J: Regarding your other question, "Am I going to continue doing research?" The answer is: absolutely. In fact, this might be the place to give you a perspective on my thinking about the NIH and about biomedical research in general. I first want to acknowledge the biomedical community and the scientific community as a whole for their commitment to serving humanity and helping the world to be a better place. They have been using the best tools that they felt were available to address the health problems that society is faced with. I'd also like to acknowledge the NIH. The support that NIH provides for biomedical research reflects our government's very deep commitment to improving and protecting the health of the nation. I feel that the NIH's intentions are in the right place.

When I told the program officer for my grant about my decision, I first requested permission to use the grant to support my new research direction, studying Maharishi Ayur-Veda. Because that research grant was for a specific research program that had undergone a peer-review process in which a large committee of scientists had examined the scientific merit of that project, it wasn't appropriate for them to allow me to use that for the new research direction that I plan to pursue. When appropriate, I will submit a grant application for research on Maharishi Ayur-Veda for peer-review.

I'm also currently the holder of a Research Career Development Award from the NIH. The response of the program administrator for that grant was quite interesting, when I told him the story that you heard today. I didn't know what his response would be. When I finished my story there was this long pause, and then the program administrator finally broke the silence, saying "You know, I think you're the perfect person to do this kind of research. You are a strong, well-established cancer researcher. You have this foundation in the molecular biosciences. At the same time, you have a natural connection to traditional medicine. This is what you should do." So they are continuing to support me with that award, which is essentially a salary award that allows me to focus exclusively on research. There are two years of this five-year award remaining. What this means is that during these next two years I can go very deeply in this new direction. We've already had exciting preliminary results in a breast cancer model and I'll be pursuing that as well as research into other Ayur-Vedic approaches that I feel are promising for dealing with cancer and other important diseases.

R: Do you reject the concept of genetic engineering? Also, your moratorium—can you tell me why it is so important?

J: I am calling for a 50-year moratorium to allow more research to be done. My basic position is that I feel we don't know enough to safely release genetically engineered organisms into the environment, and we certainly do not know enough to carry out germ-line manipulations on humans. The environment is extremely complex. It is an extended, highly interconnected meshwork of life-forms. You alter one of these, and inevitably you will influence everything else. I have talked to many environmentally-oriented scientists about this and their feeling is that the ecosystem is so complex, that it's essentially impossible to accurately model or predict the effects that would result from any alteration.

If we couldn't predict that fluorocarbons were going to react with and deplete the ozone layer, it is much less likely that we will be able to predict what will happen when we put into the soil Klebsiella that have been genetically engineered to convert wood chips into alcohol. It turns out that they cause serious problems. If you mix these bugs with soil they destroy its fertility by competing with natural soil microbes and by generating ethanol and other substances that are toxic to plants and other microbes. If you plant wheat on such soil, the wheat will sprout, but then it falls over dead. In the soil are fungi that are important in allowing plants to take up nitrogen and other nutrients. The genetically engineered bacteria suppress the levels of these fungi more than 70%. There is evidence that they also harm other microflora. We just don't know enough, even about the effects of releasing simple organisms like bacteria into the environment. We need more information.

A good example is when a kid of 10 or 12 says, "Hey, Daddy, can I drive the car?" Driving a car looks like an exciting thing for a 12-year-old, but the father knows that the child is too small, his coordination isn't good enough and that he doesn't know the laws. Natural laws are just as important. Just as Bobby needs to mature before he learns to drive, I think that society needs to mature in its knowledge of genetic engineering before we dare apply this knowledge widely. This will take at least 50 years. Much more basic research is needed and much slower implementation is essential for any application that could harm the environment.

R: So you do not want to stop genetic research?

J: Absolutely not. I think we need to re-prioritize so that, in medical research, more emphasis is put into preventive approaches. But I am definitely not saying that everyone should stop research. That would be like trying to push a river upstream.

R: You mentioned new directions of research. What do you plan?

J: I'm going to pursue research into traditional medical systems. Our culture is one of the youngest, but there are many cultures that have well-developed traditional medical systems, and we can learn a lot from them. The oriental system is very rich; the Ayur-Vedic system from India is very systematic and well-developed. Now, because of cultural and political impacts on these societies, some of that knowledge has been muddled in the last few centuries, but the knowledge is still there, and can be restored and reformulated so that it can be applied in a modern context. This is what I want to do.

There are technologies available that the West is not even aware can be useful in improving health and fighting disease. For instance, meditation is very effective. In fact, there is an NIH grant right now supporting research by one of my colleagues investigating the use of the Transcendental Meditation technique to treat high-blood pressure and cardiovascular disease. His results indicate that this approach is extremely effective. Now, such approaches are usually considered to be alternative approaches by most, but this grant is through the NIH Heart, Lung and Blood Institute. This indicates that effective approaches from traditional medical systems can be rapidly incorporated into mainstream medicine, once efficacy is established.

R: How much money is left in your grant?

J: For the Research Career Development Award? It's enough to pay me for two years, basically. That's what that grant is for.

R: Can you tell us how much that is?

J: It's in the range of $100,000.

R: You mentioned that genetically engineered crops can lead to over-use of herbicides, and the dangers to public health. Would you elaborate?

J: It's a good question and it's one that is critical because herbicide use ties the farmer to big chemical companies. They're tied to high-tech approaches. And just as I feel that prevention is the logical way to go in medicine, the way to go in agriculture is with sustained agricultural practices. I'm not an expert in that area, but there are many texts that describe the approaches that can be taken. And I can give you some references in that area. Sustained agriculture creates self-sufficiency for the farmer which is very critical in developing nations, but it's critical in the West as well.

R: What role did Maharishi play in your decision to do this and did he promise you any support?

J: Good question. Maharishi Mahesh Yogi is the founder of Maharishi International University. He is also the founder of the Transcendental Meditation and TM-Sidhi program. I referred earlier to conducting research on the Maharishi Ayur-Veda approach to health. We call it Maharishi Ayur-Veda for the same reason that we use terms like "Watson-Crick base-pairing" and "Newton's laws of gravity." Maharishi was central in bringing out and restoring knowledge of Ayur-Veda and in stimulating the extensive research that has been done on this topic, just as Watson and Crick have been central in bringing out the whole of genetic technology and Newton was central in understanding gravity.

But in terms of the decision to stop my DNA research, it was completely from my side. It's something that I have come to based on my perception of the direction that the field is going. In terms of support, no one has promised me support. I hope that the media will point out that John Fagan is looking for research support from individuals who are concerned about the future health and welfare of humanity and the environment. If support can be found, we can move very quickly to establish the utility of Ayur-Vedic approaches.

There is one other point that I would like to make in response to your question. I was able to take this stand because I am in a unique educational environment. I mentioned before the unholy alliance between science and industry, and pointed out that part of the problem is that the same people who are doing the research are involved in biotech businesses and are involved in establishing and administering regulatory policy for biotech.

I was talking with someone the other day, who described a recombinant DNA advisory meeting that they had attended at the NIH. At one point, more than half of the people on the committee had to excuse themselves from discussing a particular application because they all had financial interests associated with that application. It's a very small community.

There is great pressure on faculty to follow the party line, because the universities are all linked to the biotech industry in a very deep way. Not only have universities invested a great deal in biotech, but many faculty have started their own companies, and, of course the universities have shares in these companies. I can give you the names and phone numbers of scientists who have experienced serious professional problems when they spoke up on biotech safety issues, even in a very quiet way.

So the point I'd like to make is that the reason that I could take this action today is because MIU is sufficiently idealistic and also sufficiently unfettered with industrial and business ties to biotech, so that the administration could consider my concerns objectively. The outcome of that assessment was to strongly support my decision. They could look at the pros and cons of the dangers and say, "Fagan, I think that's the right thing for you to do."

R: Would you explain why your research on cancer susceptibility genes would have dangerous applications through genetic engineering?

J: A cancer susceptibility gene might influence one's susceptibility to cancer by 15–20%. Now this information is not likely to be used for treatment, either now or in the future. I conclude this from considering what we know about oncogenes and tumor suppressor genes. That information is unlikely to ever be useful in treating cancer. Once the oncogenic process is triggered and a tumor is formed, correcting the oncogene, even in every tumor cell, will not necessarily make that tumor go away. The many cellular changes subsequent to activating an oncogene will not necessarily be reversed by correcting a mutation in that oncogene. There isn't a lot that can be done in terms of treatment. So we're left with prevention through genetic engineering.

There are two strategies for prevention. One is germ-line; the other is somatic cell gene manipulation. Now let's consider a carcinogen that has the ability to hit every cell in the body. To somatically engineer protection means that you would have to re-engineer every cell in the body. Not practical. Well, okay, let's talk about a carcinogen that only acts on the liver. You still have a trillion cells or more to re-engineer, just in the liver. The probability of missing a few is very significant, and if you do miss a few, the treatment will be ineffective. So we see that somatic cell gene manipulations are very cumbersome, technically, and very unlikely to be successful for cancer.

We're left with germ-line gene therapy: You start with a fertilized embryo or egg, you re-engineer the DNA in that single cell. Then, when it develops, every cell in the organism will carry the genetic corrections that you carried out on the embryo. But because our ability to manipulate the genome is very crude, this procedure carries with it huge risks. With current technology the risks are so great that no one is likely to try germ-line manipulations on human embryos. Present methods are unable to insert a gene into a defined location in the genome; it is inserted randomly in most protocols.

Thus, every genetic manipulation is a potentially mutagenic event. You may insert that gene into the middle of some other gene that's important for a critical step in development, thereby causing a birth defect. You could insert it into the middle of a tumor suppressor gene, thereby creating an individual who is more susceptible to cancer, instead of less.

So the dangers are serious and even if we get to the point where we can control the insertion process more effectively, there will always be the risk of mistakes, accidents. What that means is that, germ-line manipulations will never be completely safe. We generally think of surgery as being a very invasive medical procedure, something that is not done lightly. Genetic manipulations are even more invasive. Genetic engineering is essentially surgery on something very intimate and delicate, the blueprint of our physiology.

R: Have other scientists in the community expressed any support, or are they going to follow your lead as far as returning money or making a public statement against genetic engineering?

J: Follow my lead to return the money? Nobody's volunteered. I've discussed it with some people, but they haven't acted. But I do have significant support from others in the community. If you look in the press packet, there's a page of statements of support, and there are many others out there who support this position.

APPENDIX II
DETAILED DESCRIPTION
OF FIGURE 13

Figure 13: The Constitution of the Universe

In recent decades, modern science has systematically revealed deeper layers of order in nature, from the atomic, to the nuclear and subnuclear levels of nature's functioning. This progressive exploration has culminated in the recent discovery of the unified field of all the laws of nature—the ultimate source of order in the universe.

Similarly, the ancient Vedic wisdom, understood and reformulated in this scientific age by His Holiness Maharishi Mahesh Yogi in his Vedic Science and Technology, identifies a single, universal source of all orderliness in nature, and a practical, scientifically validated procedure to apply this most fundamental and powerful level of Natural Law for the benefit of humanity.

Both understandings, modern and ancient, locate the unified source of nature's perfect order in a single, self-interacting field of intelligence at the foundation of all the laws of nature. This field sequentially creates, from within itself, all the diverse laws of nature governing life at every level of the manifest universe.

The self-interacting dynamics of this unified field constitutes the most basic level of nature's dynamics, and is governed by its own set of fundamental laws. Just as the constitution of a nation represents the most fundamental level of national law and the basis of all the laws governing the nation, the laws governing the self-interacting dynamics of the unified field represent the most fundamental level of Natural Law and the basis of all known laws of nature. The laws governing the self-interacting dynamics of the unified field can therefore be called the Constitution of the Universe—the eternal, non-changing basis of Natural Law and the ultimate source of the order and harmony displayed throughout creation.

In the unified quantum field theories of modern physics, the precise mathematical form of these fundamental laws is found in the Lagrangian of the superstring and the $N=1$ supergravity theories. In Maharishi's Vedic Science, these same fundamental laws—the Constitution of the Universe—are found in the eternal, self-referral dynamics of consciousness knowing itself. This eternal dynamics is embodied in the very structure of the sounds of the Rik Veda, the most fundamental aspect of the Vedic literature.

This chart reveals that the two descriptions of the self-interacting dynamics of the unified field—the Constitution of the Universe—provided by both modern science and Maharishi's Vedic Science are identical, and that these two great traditions of knowledge, objective and subjective—modern and ancient—uphold one another and together rejoice in providing for humanity the basic and timely knowledge of Natural Law which alone is competent to eliminate all problems and to raise the quality of life in society to the level of Heaven on Earth.

Figure 13a: Extended Discussion of the Vedic Description of the Constitution of the Universe

First, the chart displays, from the standpoint of Maharishi's Vedic Science, the self-interacting dynamics of the unified field—the Constitution of the Universe—in the structure of the Rik Veda Samhita, as brought to light by Maharishi's Apaurusheya Bhashya of the Veda (Maharishi's Commentary of Rik Veda).

According to Maharishi's Apaurusheya Bhashya, the structure of the Veda provides its own commentary—a commentary which is contained in the sequential unfoldment of the Veda itself in its various stages of expression. The knowledge of the total Veda—the complete dynamics of the unified field of consciousness and the mechanics of symmetry breaking through which the unified field sequentially creates the manifest universe—is contained in the first sukta of the Rik Veda, which is presented in Figure 13A.

The precise sequence of sounds is highly significant; it is in the sequential progression of sound and silence that the true meaning and content of the Veda reside—not on the level of intellectual meanings ascribed to the Veda in the various translations.

The complete knowledge of the Veda contained in the first sukta (stanza) is also found in the first richa (verse)—the first twenty-four syllables of the first sukta (stanza 1). This complete knowledge is again contained in the first pada, or first eight syllables of the first richa, and is also found in the first syllable of the Veda, 'AK', which contains the total dynamics of consciousness knowing itself.

According to Maharishi's Apaurusheya Bhashya of the Veda, 'AK' describes the collapse of fullness of consciousness (A) within itself to its own point value (K). This collapse, which represents the eternal dynamics of consciousness knowing itself, occurs in eight successive stages. In the next stage of unfoldment of the Veda, these eight stages of collapse are separately elaborated in the eight syllables of the first pada, which emerges from, and provides a further commentary on, the first syllable of Rik Veda, 'AK'. These eight syllables correspond to the eight 'Prakritis' (Ahamkar, etc.) or eight fundamental qualities of intelligence which constitute the divided nature of pure consciousness.

The first line, or 'richa,' of the first sukta, comprising 24 syllables, provides a further commentary on the first pada (phrase of eight syllables): the eight-syllable structure of the first pada now appears three times. The first pada expresses the eight Prakritis (fundamental qualities of intelligence) with respect to the knower or 'Rishi' quality of pure consciousness. The second pada expresses the eight Prakritis with respect to the process of knowing or 'Devata' (dynamism) quality of pure consciousness. The third pada expresses the eight Prakritis with respect to the known or 'Chhandas' quality of pure consciousness. Together, these three padas comprise the first richa (verse) of the Veda, which represents another complete stage in the sequential unfoldment of knowledge—i.e., one complete version of the Constitution of the Universe.

The subsequent eight lines complete the remainder of the first sukta—the next stage of sequential unfoldment of knowledge in the Veda. These eight lines consist of 24 padas (phrases), comprising 8x24=192 syllables. According to Maharishi's Apaurusheya Bhashya (Maharishi's Commentary of Rik Veda), these 24 padas of eight syllables elaborate the unmanifest, eight-fold structure of the 24 gaps between the syllables of the first richa (verse). Each line consists of three padas which, as in the first richa, respectively present the structure of self-interaction with respect to the Rishi (observer quality), Devata (dynamism quality—process of observation), and Chhandas (observed quality) qualities of pure consciousness. Ultimately, in subsequent stages of unfoldment, these 192 syllables of the first sukta (stanza) get elaborated in the 192 suktas that comprise the first mandala (circular cyclical eternal structure) of the Rik Veda, which in turn gives rise to the rest of the Veda and the entire Vedic literature.

This perfectly orderly, eternal structure of knowledge—the Veda—has been preserved over thousands of years in the Vedic tradition of India. The complete knowledge of the Veda and its profound significance for life has been revived and understood in a scientific framework by Maharishi Mahesh Yogi in his Vedic Science and Technology.

Figure 13b: Extended Discussion of the Modern Physics Description of the Constitution of the Universe

It is a highly significant feature of our scientific age that this complete knowledge of Natural Law provided by Maharishi's Vedic Science is now open to scientific confirmation through the unified quantum field theories of modern physics. Indeed, we see below that precisely this same mathematical structure of sequential unfoldment of the self-interacting dynamics of Natural Law is now available in the mathematical structure of the unified field found in the Lagrangian of the superstring, which represents the most complete mathematical expression of the detailed structure and dynamics of the unified field.

As with the structure of Veda, the Lagrangian of the superstring can be seen in various stages of unfoldment. The most compact presentation of the string dynamics is provided by the ten-dimensional formulation of the heterotic string ($\mathcal{L}(10)$). In addition to purely bosonic modes associated with the abstract space-time arena in which the string moves, the mathematics reveals precisely eight fundamental fermionic degrees of freedom intrinsic to the string itself—the unique solution allowed by mathematical and quantum-mechanical consistency of the theory. These eight fundamental modes of the string correspond, in Vedic terminology, to the eight Prakritis—the fundamental qualities of the unified field of consciousness. As in the structure of the Veda, these eight fundamental modes admit three interpretations corresponding to Rishi (observer quality), Devata (dynamism quality), and Chhandas (observed quality), consistent with the quantum-mechanical structure of the theory: (1) Each of the fields $\psi^{i=1...8}$ above defines a particular perspective in abstract Hilbert space (Rishi), i.e., their eigen vectors form a basis in Hilbert space which can be used to expand and interpret any other state. (2) Each of the fields $\psi^{i=1...8}$ is an operator (Devata), which creates and destroys specific states in Hilbert space. (3) Each of the symbols $y^{i=1...8}$ also denotes a particular vibrational mode or state (Chhandas) in Hilbert space, created or destroyed by its corresponding operator. With these three interpretations afforded by the quantum principle, one obtains the identical 3x8=24-fold structure corresponding to the first richa (verse) of the Rik Veda.

The next stage in the sequential elaboration of the self-interacting dynamics of the unified field is found in the free-fermionic formulation of the string in four dimensions ($\mathcal{L}(4)$). In this more expressed formalism, all bosonic degrees of freedom associated with the original, abstract space-time arena are fermionized, except for two right-moving and two left-moving coordinates needed to account for the four-dimensional structure of classical space-time geometry. This yields precisely 64 fermionic degrees of freedom intrinsic to the string itself [i.e., the 20 left movers ($\bar{\psi}^{1,2}$, χ^i, y^i, ω^i; $i=1...6$) and 44 right movers (ψ^i, ω^i, y^j, $\bar{\psi}^j$, $\bar{\eta}^k$, η^k, ϕ^m, $\bar{\phi}^m$; $i=1...6$, $j=1...5$, $k=1...3$, $m=1...8$) shown in Figure 13B]. When these 64 string fields are interpreted with respect to Hilbert space, operators, and states, this gives 3x64=192 fundamental expressions of Natural Law at this level of description of the Constitution of the Universe—in precise correspondence with the first sukta of the Rik Veda.

This precise mathematical correspondence between the descriptions of the detailed structure of Natural Law provided by modern science and by Maharishi's Vedic Science—both on the verbal level of nature's language and on the mathematical level of symbols—gives great confidence that the knowledge of the most fundamental level of Natural Law, the Constitution of the Universe, is now fully available to humanity.

This text appeared as an announcement in the international press including: *The International Herald Tribune* (International), 8 January 1992; *The Financial Times* (Great Britain), 8 January 1992; *The Financial Times* (International), 8 January 1992; *The Ottawa Citizen* (Canada), 30 January 1992; *The Globe & Mail* (Canada), 30 January 1992; *The Wall Street Journal*—Europe (Europe), 8 January 1992; *The Asian Wall Street Journal* (Asia), 10 January 1992; *The Wall Street Journal* (USA), 6 January 1992; and *The Washington Post* (USA), 9 January 1992.